Land Use
and the Causes
of Global Warming

TO WILLIAM, EILEEN, JACK AND MARY
OUR PARENTS

Land Use
and the Causes
of Global Warming

W. Neil Adger and Katrina Brown

Centre for Social and Economic Research on the Global Environment
University of East Anglia and University College London, UK

JOHN WILEY & SONS
Chichester · New York · Brisbane · Toronto · Singapore

Published in 1994 by John Wiley & Sons Ltd,
Baffins Lane, Chichester,
West Sussex PO19 1UD, England
Telephone National Chichester (0243) 779777
International (+ 44) (243) 779777

Other Wiley Editorial Offices

John Wiley & Sons, Inc., 605 Third Avenue,
New York, NY 10158–0012, USA

Jacaranda Wiley Ltd, 33 Park Road, Milton,
Queensland 4064, Australia

John Wiley & Sons (Canada) Ltd, 22 Worcester Road,
Rexdale, Ontario M9W 1L1, Canada

John Wiley & Sons (SEA) Pte Ltd, 37 Jalan Pemimpin #05–04,
Block B, Union Industrial Building, Singapore 2057

Library of Congress Cataloging-in-Publication Data

Adger, W. Neil.
   Land use and the causes of global warming / W. Neil Adger and Katrina Brown.
   p. cm.
   Includes bibliographical references and index.
   ISBN 0-471-94885-3
   1. Land use — Environmental aspects. 2. Global warming. 3. Greenhouse effect, Atmospheric. 4.
Forests and forestry — Environmental aspects. I. Brown, Katrina. II. Title.
HD156.A34 1994
333.73'13—dc20   94-15145
                                                                                 CIP

British Library Cataloguing in Publication Data

A catalogue record for this book is available from the British Library

ISBN 0-471-94885-3

Typeset in 10/12pt Times by Vision Typesetting, Manchester
Printed and bound in Great Britain by Bookcraft (Bath) Ltd

Contents

Preface

The central thesis of this book is that land use change is driven by social and economic forces, and although land use change may be a significant contributor to the emission of greenhouse gases, land use decisions cannot be appraised on this basis alone. The impetus for this book came initially from the examination of land use changes in the UK. The fluxes of greenhouse gases associated with these changes are only one of a number of important ecological and environmental impacts. The driving forces in land use change are social and economic, hence the requirement to marry two distinct ways of looking at the world into the analysis of the link between land use and global warming. The idea of using afforestation to offset the greenhouse effect has been widely publicised: the forestry lobby in particular has seized upon the opportunity to promote all forestry as benign to the global environment through being the 'lungs of the world', a phrase first applied to the tropical rainforest regions only. We do not agree with this simple representation of forestry, and indeed our UK research has shown the opposite: that afforestation can cause a net loss of carbon to the atmosphere, where previously semi-natural or mature forest is cleared or peatbogs are drained to make way for new planting.

We extend this analysis in several ways by showing the relative size of the problem in a UK context and by examining land use change and the emission of greenhouse gases at a global level. Social science analysis requires that the costs, impacts and benefits of different strategies are addressed from different perspectives because the impacts of global warming will occur in the future and will be geographically and socially skewed. The costs of undertaking precautionary abatement action are also skewed, and it is this theme which is developed in this book. We draw upon examples of land use change which are global and regional, in addition to the UK case study.

The most significant political manifestation of concern over the global warming issue is the UN Framework Convention on Climate Change which entered into force in March 1994. The Convention initiated an ongoing process of negotiations, reporting and target setting with the objective of the stabilisation of global emissions, both through emission reduction and sink enhancement. The task of this book is to inform those planning and researching in the area of global warming of some of the links with land use and land use change. In particular, the book examines the type of policies which could be implemented in the land use sector, their likely impacts on greenhouse gas emissions, and the main factors and constraints influencing land use decisions. Our audience is therefore researchers, academics in the fields of global environmental change, environmental economics and management, agriculture and forestry, and administrators and scientists who are close to the policy-making process.

We would like to thank many people who have contributed to the production of this volume. The directors of the Centre for Social and Economic Research on the Global Environment (CSERGE) at the University of East Anglia and University College London, Kerry Turner, Tim O'Riordan and David Pearce encouraged, advised and enabled us to undertake this task. CSERGE is core funded by the Economic and Social Research Council as part of a major programme of research on global environmental issues. Our colleagues at the Countryside Change Unit and the Department of Agricultural and Environmental Science at the University of Newcastle upon Tyne, Robert Shiel and especially Martin Whitby collaborated on early work and contributed greatly to the development of our ideas. The Countryside Change Unit was also funded by the ESRC. Dominic Moran and Francis Mudge were involved in collaborative research on particular issues. We are grateful to Susan Subak, Andrew Jordan, Mick Kelly, Sam Fankhauser and Dick Cobb who read drafts of various parts of the book and advised on issues from economics to biogeochemical cycles and climate change. They should in no way be held liable for mistakes or opinions expressed in the volume. This responsibility rests with us, as authors, alone. Lastly, we would like to thank Pauline Seeley, Rosie Cullington and Alex Howe for typing parts of the manuscript and to Philip Judge for graphics.

Neil Adger
Katrina Brown
Norwich
July 1994

PART I

INTRODUCTION AND OVERVIEW

CHAPTER 1

Land Use and Global Environmental Change: A Social Science Perspective

1.1 INTRODUCTION

Global warming, caused by the increase in atmospheric concentration of greenhouse gases, has its roots in anthropogenic activity. The natural greenhouse effect keeps the earth warmer than it would otherwise be, but human activities enhance this greenhouse effect, resulting, on average, in additional warming of the earth's surface. The primary causal anthropogenic activity is the burning of fossil fuel for energy, but land use and land use change are also considerable sources of the increased concentrations. These land use causes of the enhanced greenhouse effect leading to global warming and other impacts, are the foci of this book. Land use is determined by environmental factors such as soil, climate and vegetation, but it is also combined with the other classic factors of production, labour and capital, in economic production. Thus land undisturbed by human intervention at any time in recent history is uncommon, and increasingly the human uses of land as a factor of production fundamentally determine land use and land use change.

The aim of this volume is to identify the physical processes relating land cover and use to the global warming system and to analyse the economic and other factors driving land use and land use change to show the context within which action on the global warming problem can feasibly be taken. Although the climate system is inherently variable, the predicted rates of change of climate will overwhelm the migratory and evolutionary abilities within ecosystems and could lead to further feedbacks into the climate change process. As land is inherently necessary for the provision of food and fibre, the climate change impacts have huge ramifications for the whole population of the planet, no matter what technological advances are made. This takes only the human-centred view of the problem. The survival of numerous other species is also at risk.

The role of land in causing enhancement of the greenhouse effect is a crucial political question. Land is the primary economic resource and also the resource which geographically defines spheres of power and influence. These interrelated factors were recognised in the Framework Convention on Climate Change signed by over 160 countries at the UN Conference on Environment and Development in 1992. The Convention accepts that climate change will impact on different countries with different severity and that each country has a common but differentiated responsibility for present causes and future action to reduce emissions and enhance sinks. The economically disadvantaged countries of the South are less well adapted to coping with climate change than the industrialised North. They also have less leeway in abating emissions associated with land use, and have a moral case that the industrialised countries which deforested centuries ago and industrialised first should take the greatest steps to avoiding the potentially catastrophic impacts of climate change. The reverse of this argument, that all avenues for greenhouse gas abatement should be explored, and that least-cost actions should be undertaken, does not persuade developing countries unless it is accompanied by incentives for action.

The major determinants of land use and land use change are physical, climatic and demographic factors, levels of poverty, and the economic and institutional structures of resource use. Economic factors include the demand for primary commodities such as agricultural products and forest products. The institutional factors which underpin these systems of exchange also determine the systems of use of the resources (e.g. Kates and Haarmann, 1992).

In this introductory chapter the role of land use in the global cycles of the main greenhouse gases is explained, specifically as sources and sinks of carbon dioxide (CO_2), methane (CH_4) and nitrous oxide (N_2O). The impacts of these emissions are global warming and related changes in the biosphere; the specific impacts on agriculture and other ecosystems are summarised. These impacts are uncertain and have not been directly observed to date, so any social science analysis necessarily focuses on both the causes of the problem and on the nature of decision-making. The phenomenon of anthropogenic interference with the atmosphere, and the resulting global warming, is fundamentally *a global commons issue*. No group or individual retains the property rights to the atmosphere and those who emit greenhouse gases do not bear the consequences of their actions. The chapter sets out the economic framework of the problem of global warming. This approach is limited, however, by the international political context with respect to land use sources and sinks. Issues of present equity and historic responsibility for global warming; the relative cost of abatement in developed and developing countries; and the issue of international transfers all demonstrate that the narrow economically efficient solutions to global warming may well be infeasible.

1.2   THE CARBON CYCLE AND THE PRINCIPAL CAUSES OF GLOBAL WARMING

The greenhouse effect is the result of the presence in the atmosphere of gases such as carbon dioxide, the so-called greenhouse gases, which absorb outgoing terrestrial radiation while permitting incoming solar radiation to pass through the atmosphere relatively unhindered. Although the basic heat-trapping mechanism keeps the atmosphere of the earth relatively warm compared to other planets and atmospheres in the solar system, the enhancement of the greenhouse effect due to the increasing atmospheric concentration of greenhouse gases is a cause for serious concern. The reasons for this concern are outlined below, namely the potential for global climate change, sea level rise and related impacts.

The major greenhouse gases associated with land use are CO_2, CH_4 and N_2O. Some other trace gases implicated in the greenhouse effect have no biological sinks and few land use related sources. These are the halocarbons, including the chlorofluorocarbons, and ozone. Table 1.1 shows some important characteristics of these substances and a summary of their sources and sinks. The order of importance of the gases in relation to their contribution to global warming decreases left to right as CO_2 accounts for at least 50 percent of the enhancement of the greenhouse effect at present, however estimated. This section reviews the sources, sinks and relative importance of these gases in enhancing the greenhouse effect and the relationship with biological sources and sinks.

Carbon dioxide

Carbon dioxide is currently increasing at 0.5 percent per annum in the atmosphere, and now constitutes approximately 355 parts per million by volume (ppmv) compared to 280 ppmv in pre-industrial times (see Boden et al., 1991; Watson et al., 1992). Estimates of the global biomass pool range from 550 to 830 billion tonnes of carbon (bt C) (1 billion = 10^9) (Bouwman, 1990a), with the other major pools being the atmosphere (720 bt C), oceans (38 000 bt C) and in fossil fuel reserves (6000 bt C). The interactions of these pools are shown schematically in Figure 1.1. Net changes in these fluxes between the major pools have been summarised by the reports of the Intergovernmental Panel on Climate Change (IPCC) and are shown in Table 1.2. The IPCC (Houghton et al., 1990) estimates emissions from deforestation at 1.6 bt C (1 t C = 3.67 t CO_2), the central estimate of fossil fuel emission rates being 5.4 bt C. The estimates of emissions from natural sources are, however, subject to great variation caused mainly by ranges in estimates of tropical deforestation.

The accumulation in the atmosphere is the net impact of the interactions of land, ocean and atmosphere illustrated in Figure 1.1. Over the long term, about 40 percent of anthropogenic emissions has remained in the atmosphere (Wuebbles and Edmonds, 1991). On a yearly basis at present, the estimates presented by the IPCC, derived from observational and model data, suggest a net imbalance in the budget of 1.6 bt C per year in the 1980s; the present rate of increase of the atmospheric pool of CO_2 is less than half of that expected given the estimates of the magnitude of the major sources and sinks. There appears, then, to be a

Table 1.1 Atmospheric concentrations, increase, residence time, sources and sinks for the major greenhouse gases, and their contribution to global warming

	CO₂	CH₄	N₂O	O₃	CFCs
Residence time (year)	100	8–12	100–200	0.1–0.3	65–110
Annual increase (%)	0.5	1.0	0.2–0.3	2.0	3.0
1990/1991 concentration	355 ppmv	1.72 ppmv	310 ppbv	na	0.28–0.48 ppbv
Radiative absorption per ppm of increase	1	32	150	2000	>10 000
Contribution (%) to global warming	50	19	4	8	15
Total source	6.5–7.5 bt C	400–640 mt CH₄	11–17 mt N	—	—
% biotic	20–30	70–90	90–100	—	—
Major sources	Fossil fuel use (5.4 bt), deforestation and shifting cultivation (1–2 bt)	Rice paddies (60–140 mt), wetlands (40–160 mt), ruminants (65–100 mt), termites (10–100 mt), landfill sites (30–70 mt), oceans/lakes (15–35 mt), biomass burning (50–100 mt), fossil (50–95 mt) (coal mining + gas exploitation)	Cultivated soils (3 mt), natural soils, fossil fuel	Atmospheric	Industrially manufactured
Sinks	Atmospheric accumulation (3.5 bt), oceans (<1 bt), biosphere, charcoal formation	Atmospheric accumulation (30 mt), soil oxidation (32 mt), atmospheric chemistry (420–520 mt)	Atmospheric accumulation (2.8), atmospheric chemistry (10.5) Soils	Atmospheric accumulation, atmospheric chemistry	Atmospheric accumulation, atmospheric chemistry

Sources: Updated from Bouwman (1990b), Shine et al. (1990) and Watson et al. (1992).

Notes: ppmv = parts per million by volume, ppbv = parts per billion by volume, na = not available.

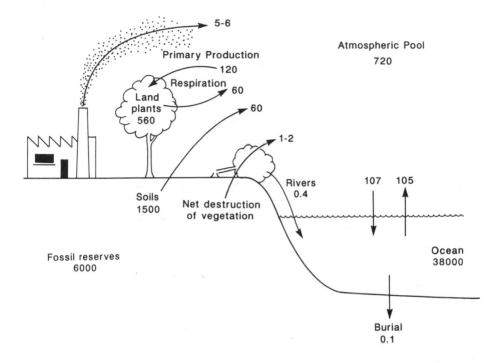

Figure 1.1 The global carbon cycle.
Notes: figures in billion tonnes of carbon. Adapted from Schlesinger (1991)

'missing' carbon sink. How this anomaly is resolved, and where this sink resides, has important implications for climate change policy and especially for policy towards the biological sources and sinks.

Missing carbon sink

The missing sink has been variously postulated to be in temperate forests, soils or in the oceanic carbon pool (see Tans et al., 1990; Kauppi et al., 1992; Sedjo, 1992; Tate, 1992; Duxbury and Mosier, 1993). In Table 1.2, the missing sink, the net imbalance in the carbon budget, is estimated as 1.6 ± 1.4 billion tonnes (bt) C year^{-1}. One possible resolution of the missing sink problem is suggested by Tans et al. (1990). They model the terrestrial and oceanic carbon sinks by latitude, based on the observed north–south atmospheric CO_2 concentration gradient. They conclude that terrestrial ecosystems are a considerably greater sink than the oceans and that there must be an extra terrestrial sink located in temperate latitudes to balance the carbon budget. This sink is in the range of 2.0–3.4 bt C year^{-1}. Sedjo (1992) presents evidence which suggests that this temperate carbon sink is almost certainly in forests. Since forests contain possibly 85 percent of the

Table 1.2　1980s Global carbon fluxes (billion tonnes carbon)

Emissions from fossil fuels	5.4 ± 0.5
Emissions from deforestation and land use	1.6 ± 1.0
Accumulation in the atmosphere	3.4 ± 0.2
Uptake by the ocean	2.0 ± 0.8
Net imbalance	1.6 ± 1.4

Source: IPCC 1990 report (Watson et al., 1990, p.13).

above-ground biomass, carbon changes in forest stock will have a large influence on the carbon pool. This is recognised in the context of tropical forests where deforestation is the principal source of emissions (Houghton et al., 1991). However, Sedjo's (1992) contribution is to show from cross-country forest assessments that the land area under forest and the forest stock of biomass has been expanding in the northern hemisphere. In some areas of the northern hemisphere this has occurred for over a century, but it is predominantly in the past several decades through both afforestation programmes and through reforestation by natural regeneration. This is backed up by a recent FAO assessment (Korotkov and Peck, 1993) which shows an increase of forested land in Europe of 1.9 million ha in the 1980–1990 decade, as well as felling rates of around only 70 percent of net annual incremental growth for Europe. Sedjo's estimate of annual carbon sequestration of the temperate forests of North America, Europe and the former USSR is around 0.7 bt C year^{-1} and this largely explains the missing sink, as hypothesised by Detwiler and Hall (1988). Unfortunately, this still does not match or explain the modelled missing sink of 2.0–3.4 bt C of Tans et al. (1990).

This _ ,cent conclusion concurs with other studies pointing to the missing sink ͻeing in temperate regions. Studies which assume that mature forests or other ecosystems in steady state have zero net impact on the carbon balance are naïve, since systems are predominantly recovering from past disturbance which, although these may have been decades in the past, still results in considerable active carbon sequestration across large land areas previously thought to have been inactive. A further factor is the fertilisation effect associated with atmospheric CO_2 increases which may promote primary productivity. CO_2 fertilisation is frequently cited as a likely cause of changes in net primary productivity of vegetation and hence of the magnitude of the terrestrial carbon sink (Melillo et al., 1993; Taylor, 1993). There may also be a fertilisation effect due to the increased presence of nitrogen in forest ecosystems. These factors — rising forest stocks in temperate regions, the impact of past disturbance in promoting carbon sequestration in mature vegetation, and fertilisation effects — suggest that the conservation of present forests is of even greater importance (see Brown et al., 1992 for a summary).

However, the conclusion of Sedjo (1992) and of Kauppi et al. (1992) of the existence of the missing sink in terrestrial biota of the northern hemisphere is severely questioned in a paper by Houghton (1993a). His argument is based on a re-estimation of net carbon storage throughout the cycle of forest growth and harvest. Houghton takes issue with the findings of Sedjo and Kauppi et al. on the basis that the decomposition of litter after harvesting and the decomposition of forest products releases a much greater proportion of carbon than previously estimated.

All of these disputes show that a definitive model of the carbon cycle is not an immediate prospect. Indeed, there are other factors which should be taken into account, such as the impact of climate change itself. It has been proposed that the missing sink can be accounted for by an increase in the net primary production of vegetation throughout the terrestrial biota as a result of observed temperature and precipitation change in the last half century (Dai and Fung, 1993). Dai and Fung estimate a cumulative carbon sink of 20 ± 5 bt C for the period 1950–1984, based exclusively on temperature and precipitation change and ignoring CO_2 fertilisation.

Whatever the global assessment, it is clear that the missing sink is of critical importance for policy prescriptions. The failure of researchers to provide a full explanation of the global carbon cycle leads directly to problems in the analysis of future concentrations of CO_2, and hence climate change impacts (Edmonds, 1992). In modelling past and future concentration changes, assumptions have to be made regarding the 'location' of the missing sink. It may be assumed that historical estimates of land use emissions are in error; these may be 'corrected' to sidestep the missing sink problem. It may be assumed that CO_2 fertilisation accounts for the discrepancy in the budget. If these models are used for policy analysis, these assumptions will affect the end result, by altering the relative importance of the other sources and sinks. Further, the treatment of the missing sink creates problems in the estimation of the global warming potential (GWP) associated with each greenhouse gas. The GWPs for all gases are stated in terms of carbon dioxide equivalent emissions. But the assumptions used to account for the missing sink, whether they affect the relative importance of land use emissions or any other component, mean that the estimation of the radiative forcing of carbon dioxide (the numeraire in the equation of GWPs) may be biased, possibly up to 15 percent (Edmonds, 1992; Wuebbles et al., 1992).

Methane

The atmospheric concentration of CH_4 has been rising steadily in the last 300 years. The current concentration of 1720 ppbv corresponds to an atmospheric reservoir of around 4900 mt CH_4 which is increasing by around 30 mt CH_4 per year (see Table 1.3). There has been a marked acceleration in the atmospheric concentration in the last 100 years in a manner highly correlated with global human population (Watson et al., 1990, p.19; Bartlett and Harriss, 1993). However atmospheric concentrations have only systematically been measured for the last 30 years, the results suggesting that concentrations have increased roughly 1 percent per year over that period, with a small decrease in the rate in the late 1980s.

CH$_4$ constitutes 8–15 percent of total greenhouse gas production, depending on how the relative global warming potential of CH$_4$ to CO$_2$ is determined. Around two-thirds of these emissions are related to human activities, and one-third natural sources. Data on the global methane cycle are not accurate, and are derived from calculations of the atmospheric concentrations allocated between the different sources, given the best experimental data (Cicerone and Oremland, 1988; Khalil and Rasmussen, 1990b). Emission rates from natural sources and from human activity, such as from biomass burning and from rice cultivation for example, are extremely controversial (Crutzen, 1991; Levine, 1990; Chapter 4).

Table 1.3 shows the estimates of global emissions from the 1992 IPCC assessment. The major sink is the reaction of CH$_4$ with the hydroxyl radical (OH $\cdot$) in the atmosphere. The other sink is soil, though it is suggested that soil uptake may be decreasing globally with changing land use and fertiliser application. Unlike CO$_2$, the sources and sinks of CH$_4$ in Table 1.3 have all been identified, with the 30 mt CH$_4$ atmospheric increase per year being borne out by observation.

Of the sources, agriculture related emissions form almost half the anthropogenic emissions (even excluding the land use change related emissions from biomass burning). Emissions from ruminants are estimated to provide an atmospheric

Table 1.3   Estimated sources and sinks of methane (million tonnes CH$_4$ per year)

	million tonnes CH$_4$ year^{-1}	(range)
Natural sources		
Wetlands	115	(100–200)
Termites	20	(10–50)
Ocean	10	(5–20)
Freshwater	5	(1–25)
CH$_4$ hydrate	5	(0–5)
Anthropogenic sources		
Coal mining, natural gas and petroleum industry	100	(70–120)
Rice paddies	60	(20–150)
Enteric fermentation	80	(65–100)
Animal wastes	25	(20–30)
Domestic sewage treatment	25	na
Landfills	30	(20–70)
Biomass burning	40	(20–80)
Sinks		
Atmospheric removal	470	(420–520)
Atmospheric increase	30	(15–45)
Removal by soils	32	(28–37)

Source: Watson et al. (1992, p. 35).   na = not available

source of 65–100 mt CH_4 year $^{-1}$, which is based on the estimates of emissions per animal multiplied by the global ruminant animal population. Many countries with low per capita incomes are among the largest emitters through having large ruminant populations. Reductions in herd sizes may be difficult to obtain due to the subsistence nature and cultural significance of livestock herding. Table 1.3 shows that the handling of animal wastes adds almost one-third to the source of domestic animal emissions. This is largely in industrialised countries where intensive management leads to an increased source with the same herd size. The issues of reducing methane emissions from livestock and from rice growing are examined in detail later in the book.

Nitrous Oxide

Atmospheric nitrous oxide emissions are currently rising at a rate of approximately 0.8 ppbv per year, so the 1990 concentration estimated in Table 1.1 of 310 ppbv is now likely to be 310–320 ppbv (Watson et al., 1992). The atmospheric N_2O cycle and in particular the capacities of the various sources and sinks are poorly understood. The principal sink for N_2O is destruction by ultraviolet light in the stratosphere, which accounts for its long atmospheric residence time of approximately 150 years. Of the estimated 14 mt of N_2O emitted to the atmosphere each year, only a small fraction is now thought to come from fossil fuel use. The rest arises from gas exchanges with the oceans, from biomass burning and deforestation, and is emitted from soils through the processes of denitrification and nitrification, both of which are associated with the application of agricultural fertilisers.

Nitrification is microbial transformation of ammonia (NH_4^+) to nitrate (NO_3^-), while denitrification is the transformation of NO_3^- to molecular nitrogen (N_2). Nitrification is dominant in aerobic soils, whereas denitrification occurs in anaerobic conditions. Factors influencing the speed of turnover of N are soil temperature, moisture, fertility, availability of organic matter, and drainage, and thus N_2O fluxes vary considerably both spatially and temporally (Bouwman, 1990a; Bouwman et al., 1993).

The rate of loss of N will depend on the timing of fertiliser application, and the type of fertiliser used. While both natural and synthetic fertilisers contribute to N_2O emissions, synthetics are assumed to be the main offenders, largely because they comprise the 'extra' nitrogenous elements added to the system (Ehrlich, 1990). Bolle et al. (1986) estimate that based on total production rates of different types of fertilisers, loss of mineral fertilisers in the form of N_2O is 0.5–2.0 percent. The same amount may be emitted from denitrification and/or nitrification of mineral fertilisers leaching from the fields into groundwater or surface freshwater ecosystems, and then emitted to the atmosphere. The total loss rate of fertiliser nitrogen then becomes 1–4 percent of that applied. Duxbury and Mosier (1993) criticise the IPCC assessment of the nitrous oxide cycle for underestimating the impacts of increased nitrogen cycling brought about by fertiliser application as well as its failure to take grassland emissions into account. Grasslands have a global area almost as large as forests, and are estimated to emit 0.05–2 mt N per year.

Nitrous oxide is an important greenhouse gas from the perspective of policy analysis because of the sources and sinks identified. Chapter 4 illustrates that changed agricultural practices can reduce emissions of one greenhouse gas, such as methane from rice paddy growing, but increase emissions from another. Numerous examples of this exist in various mitigation options. Compared to CO_2 and CH_4, N_2O is, however, a relatively minor contributor to the global warming problem.

Global warming potential of the gases

With the multiple sources of a number of greenhouse gases outlined above, it is important to be able to compare the gases on a common basis, for example to estimate individual country contributions to the greenhouse effect (Fujii, 1990; Hammond et al., 1990; World Resources Institute, 1990). Comparing emissions of the different gases is only one of a number of problems in constructing a responsibility index, others being whether historical or present emissions are calculated or whether total country or per capita emissions are the relevant unit of account. The indices of responsibility which have been constructed are highly controversial and have stimulated much debate; they are also highly sensitive to how the different gases are compared.

The contributions of the different greenhouse gases also need to be comparable to appraise least-cost strategies for emission reduction. Clearly, if the costs of reducing emissions of CO_2 and CH_4 per unit weight were identical at the margin, it would make the greatest sense to take the course of action which reduced CH_4 emissions because of the greater potency of CH_4. Indeed if the relative potencies of the various gases are known, it makes economic sense to reduce emissions of the individual gases till the marginal cost per unit of greenhouse 'potency' is equal. The role of the different gases caused much controversy in the negotiation of the international treaty on climate change as it was argued by some parties that CO_2 alone should be the focus of attention because it is the single most significant gas; because many of the sources are relatively easily documented; and because abatement options are feasible given present and near future technologies. Many of these factors are indisputable in themselves, especially the monitorability and feasibility of reduction of CO_2 emissions compared to the other gases (Subak, 1994). From an equity perspective the inclusion of all sources generally places more responsibility with developing countries who can least afford abatement (Brown and Adger, 1993). However, the effect of focusing on CO_2 alone is to apply an implicit weight of zero to the other gases (Smith, 1993) and would lead to sub-optimal solutions, given the economic argument set out above. The contributions of the major gases to the potential rate of global warming therefore have to be made comparable.

Each greenhouse gas acts in a different manner in the atmosphere to enhance the greenhouse effect: each is said to have a different radiative forcing potential. One measure of the contribution to the greenhouse effect of the various greenhouse gases, a measure of their instantaneous effect on the energy balance, is shown in the first column of Table 1.4. The radiative forcing associated with the other major greenhouse gases relative to CO_2 is greater: CO_2 has the least forcing potential when the instantaneous effect on the energy balance is considered. However, several factors make the use of this particular measure alone misleading for comparison of the gases: these are the *atmospheric residence time* of the gases, and their *indirect impact* on the atmospheric chemistry of the other gases (see Harvey, 1993 for a review). It is necessary to estimate the cumulative effect of any emission over the lifetime of that gas in the atmosphere. Methane, for example, has a much shorter residence time in the atmosphere than CO_2, but generally breaks down into CO_2 by reacting with hydroxyl radicals. Other gases, such as carbon monoxide (CO), have no direct radiative forcing properties but do contribute indirectly to the greenhouse effect through reaction with atmospheric CH_4. There is uncertainty as to how these residence times will change as the atmosphere warms in the future. The residence time of CO_2 is likely to increase as its atmospheric concentration increases and hence the global warming potentials (GWPs) of all the other gases would change. Due to this uncertainty it has been suggested that, although CO_2 is historically the most important trace gas, it may not be the best gas to use as the base for comparison, but its central role in causing the greenhouse effect and the likely policies to reduce emissions necessitate its use.

The longevity of the different gases means that if the integration of the radiative forcing over time for gases such as CO_2, a proportion of which remains in the atmosphere indefinitely, is taken to infinity then the other gases cannot be compared since the denominator of a ratio comparing CO_2 to the non-CO_2 greenhouse gas would be infinity. A time horizon over which the summation of the radiative forcing is to be taken is required. All these factors lead to the GWP measure calculated for each of the major greenhouse gases by the IPCC in their 1990 and 1992 reports (Shine et al., 1990; Isaksen et al., 1992), and reported in Table 1.4.

The GWP is defined as the 'time integrated commitment to climate forcing from the instantaneous release of 1 kg of a trace gas relative to that from 1 kg of CO_2' (Shine et al., 1990, p.58). The global warming potential then is:

$$GWP = \frac{\displaystyle\int_0^\infty f_i(t) \cdot C_i(t)\, dt}{\displaystyle\int_0^\infty f_c(t) \cdot C_c(t)\, dt}$$

Table 1.4 Relative global warming potentials of greenhouse gases compared to CO_2

Greenhouse gas	Radiative forcing relative to CO_2	Residency (years)	Relative global warming potential (direct + indirect)			Relative global warming potential (direct only)			Sign of indirect effect
			20 year basis	100 year basis	500 year basis	20 year basis	100 year basis	500 year basis	
CO_2	1	120	1	1	1	1	1	1	none
CH_4	58	10.5	60	21	9	35	11	4	+ ve
N_2O	210	132	270	290	190	260	270	170	uncertain
CFC 11	4000	55	4500	3500	1500	4500	3400	1400	– ve
CFC 12	5700	116	7100	7300	4500	7100	7100	4100	– ve

Sources: Shine et al. (1990), Isaksen et al. (1992) and Lelieveld and Crutzen(1992).

Notes:1 unit weight CO_2 = 1. The total (direct + indirect) global warming potentials shown are those reported by Shine et al. (1990) apart from those of CH_4 which are based on updated analysis of Lelieveld and Crutzen (1992). Although the indirect effects of some of the gases are uncertain, using estimates of direct GWPs only would tend to underestimate the relative contribution of non-CO_2 gases. Further, for methane the sign of the indirect effect is likely to bring the total GWP of CH_4 close to Lelieveld and Crutzen's estimate: 'Although we are not yet in a position to calculate new indirect GWPs, we can estimate the sign most likely for some compounds based on current understanding. For example, the indirect GWP for methane is positive and could be comparable in magnitude to the direct value' (Isaksen et al., 1992, p.55). The CFC measures do not account for indirect cooling via ozone layer depletion.

where

$C_i(t)$ = the concentration of gas i at time t following the emission of a unit amount at time $t = 0$,

$f_i(t)$ = the heat trapping ability (radiative forcing effect) at the same time per unit of concentration, and

$C_c(t)$ and $f_c(t)$ = corresponding quantities for CO_2.

As can be seen, if the gases have a finite lifetime then the GWP ratio can be calculated. As noted above, however, a proportion of emitted CO_2 effectively stays in the atmosphere indefinitely, meaning that for any steady emission rate, the atmospheric concentration increases indefinitely and the lower integral of the ratio is infinite. To calculate ratios of the global warming potential of non-CO_2 greenhouse gases it is therefore necessary to integrate the radiative forcing arising from an impulse input only up to a fixed time horizon rather than to infinity. The GWP so defined then depends on the selected horizon: gases with a shorter atmospheric lifetime than CO_2 will have a smaller GWP the longer the time horizon considered. The resulting GWPs for the main greenhouse gases are shown in Table 1.4.

The IPCC 1992 report updates the estimates of the GWPs presented in the 1990 report on the basis of greater understanding of the processes involved. The direct GWPs are more certain and are reported in Table 1.4, along with the likely direction of the indirect effects. In most cases, the GWPs do not change substantially between the IPCC 1990 and 1992 reports. The reported methane GWP is from the study by Lelieveld and Crutzen (1992).

Which time horizon is most appropriate for analysis? From a policy perspective, the horizon would appear to be important only for consideration of the problem at a global level. The IPCC suggests that if the likely effects of accelerated sea level rise are the focus, then a 100 year time horizon may be appropriate; whereas for the evaluation of near term effects of climate change, such as changing rainfall patterns, a shorter time horizon is more applicable. At the local or regional level, however, the decision to abate emissions of greenhouse gases, and hence the relative importance of the gases, is not relevant to mitigation strategies. Changes in local emissions are too small to have a noticeable impact on climate and the decision is thus only concerned with protection and adaptation costs to exogenous likely, or even already committed to, scenarios of climatic impacts (Adger and Fankhauser, 1993).

Although the potential impacts of climate change and of sea level rise are likely to be highly spatially skewed, and hence of interest to different nation states, strategies for overall emission reduction from these states using the most appropriate time horizon would not necessarily produce the desired solution. The implication of the IPCC recommendation is that small island states concerned about the impacts of sea level rise should reduce aggregate emissions based on a 100 year perspective, and countries with greater interest in shorter term impacts should reduce emissions on a 20 year basis. The choice of horizon is only relevant at the global scale. The most appropriate reduction strategy across the different gases must then anticipate the uncertain impacts of global warming (as well as the other consequences of emissions). As international comparisons of national greenhouse gas emissions tend to make an *implicit* judgement of the time horizon, a basis for comparison should be made *explicit* under the Conference of the parties to the Climate Change Convention.

1.3   IMPACTS OF CLIMATE CHANGE AND FEEDBACKS

Global warming has many potentially damaging impacts and a few potentially offsetting positive impacts. Although future emissions are unpredictable in so far as they depend on present and future action on climate change, it is clear that the global-mean temperature has risen by 0.45 °C ($\pm$0.15 °C) since the middle of the 19th century, though the warming has not been continuous over time or space, with two periods of rapid warming in the period 1910–1940 and the 1970s to present (see Hulme, 1993b for a review). The earth is committed to further increases if a 'business as usual' path of emissions increase is followed: approximately 0.3 °C (range 0.2–0.5 °C) per decade over the next century, with an associated rise in mean sea levels of 6 cm per decade (range 3–10 cm per decade) (Houghton et al., 1990, 1992). The reported ranges are associated with climate sensitivity uncertainties.

Improved understanding of the various processes within the climate system driving climate change has led to the identification of some factors which may increase or decrease the rate of climate change over the next century. The previously unaccounted mechanisms include:

(1) the cooling effect of sulphate aerosols in the atmosphere;
(2) halocarbons having a direct and strong warming effect in the lower troposphere but depleting stratospheric ozone and hence causing cooling in that part of the system, with the net effect of reducing their overall warming by possibly 80 percent (Ramaswamy et al., 1992);
(3) the impact of atmospheric CO_2 fertilisation on the ability of vegetation to sequester carbon from the atmosphere (Wigley, 1991).

The result of these recently quantified impacts is to reduce the best guess central estimates of sea level rise and global mean temperature for the scenarios in which no significant emission reduction is assumed (Wigley and Raper, 1992, 1993). An assessment of committed to and avoidable impacts would require the incorporation of potential emissions scenarios, dependent on human population growth, the efficiency of energy use technologies, the rate of deforestation and other factors.

The range of global warming impacts presented by Wigley and Raper (1992) due to different emission scenarios in which action is taken to reduce emissions, are of global warming by 2100 of 1.6–3.0 °C (relative to 1990) (0.1–0.25 °C per decade) and of global mean sea level rise of 35–54 cm (3–4 cm per decade). The ranges represent the uncertainty regarding future emissions, which compounds the uncertainty concerning the mechanisms of climate change and sea level. The emission reduction scenarios differ on how CFCs are phased out, human population and economic growth, and energy source supplies, and hence produce the ranges reported by Wigley and Raper. The uncertainty over future human action will always remain much higher than that over the climate change mechanisms, as scientific progress will narrow the latter range (Hulme, 1993a).

The sectors of the biosphere potentially suffering the most significant impacts have already been identified at least in general terms. Ecological systems will be affected by temperature and precipitation change and by sea level rise, resulting from the thermal expansion of the oceans and melting of continental ice. Certain plants (C3-pathway agricultural crops such as wheat, rice and potatoes) could benefit from the CO_2 fertilisation effect, though the impact on agricultural yields and forests is also a function of water availability (e.g. Shugart et al., 1986; Bazzaz and Fajer, 1992). Ecosystems in general will migrate polewards: the more rapid the climate change, the higher the probability of disruption and surprise in the ecosystem affected (Schneider, S. H., 1992). Although some species such as birds may migrate quickly, their habitats will not 'migrate' at the same rate, with past evidence of climate changes showing that forests areas become marooned in ecologically unsuitable territory (Peters, 1988, 1990; Root and Schneider, 1993). Further, with economic development, habitats and species may be lost in coastal areas as a result of the combined effects of climate-induced sea level rise and the protection of economic assets in the coastal zone (Daniels et al., 1993), unless coastal land is abandoned and semi-natural coastal habitats allowed to develop (Reid and Trexler, 1991; Turner et al., 1993).

As climate alters, the distribution of other species will be affected and this may have negative consequences. Agricultural pests will also be displaced over time, as will the incidence of disease vectors through the spread of malaria carrying mosquitoes, and other vector-borne diseases such as schistosomiasis and lymphatic filariasis, affecting the health of vulnerable sections of human populations (Haines, 1993). In addition, between 100 million and 200 million people are estimated to be subject to much increased flooding frequency with slight mean sea level changes (IPCC Response Strategies Working Group, 1992).

The costs of climate change and related phenomena are potentially catastrophic and significant enough to reduce the potential for economic development of some developing countries. It is also the case that areas vulnerable to increased rates of climate change and its impacts also tend to be vulnerable to other acute hazards. Vulnerability has both a biophysical and a socio-economic dimension and low lying areas may already be vulnerable to flooding for example (Hewitt, 1984; Kasperson et al., 1990).

The direct effects of climate change on agriculture and forestry are more uncertain than those on secondary and tertiary sector economic activities as the parameters of precipitation, windspeed and dynamic effects on soils of temperature change are unknown. Secondary impacts such as changes in the run-off of nitrate pollutants to water courses, or increased agricultural pests are too uncertain to be considered in most analyses (Smit et al., 1988). Ignoring other effects, it would seem that a CO_2-enriched atmosphere would cause increased growth in vegetation. Yet evidence suggests that even this offsetting impact may be overstated through effects on soil fertility (due to slower decomposition in a CO_2-rich environment) and through increased competition between species for CO_2 absorption (Bazzaz and Fajer, 1992).

The physical effects of climate change on agriculture do not, however, determine the relative economic welfare changes. Measures of economic welfare reflect the fact that a change in physical production relationships translates to changes in the price of commodities. A price rise, while not good if one is a buyer, is beneficial if one is a seller of a commodity, other things being equal. Economic welfare changes are then the net effect of price *and* quantity changes. Distributional issues of whether producers or consumers benefit or lose are separate issues in such an assessment.

Although research has tried to determine the physical response relationships, it is the direct economic effects of supply changes and the indirect effects of changing world agricultural commodity prices that will determine whether global warming impacts will be significant at a local scale. Attempts to quantify the economic effects by the US Environmental Protection Agency and the US Department of Agriculture have produced a wide range of results. Kane et al. (1992) show a range of net welfare effects of projected changes in yield, hence production responses in world agriculture to climate change scenarios. Net welfare change is made up of changes in producer and consumer surplus as supply or demand shifts exogenously. The estimates are based on a US Department of Agriculture trade model which describes regional production, prices and trade in agricultural commodities. The results, however, show a range between a positive net welfare effect (0.01 percent of world GDP) and a negative effect (0.47 percent of world GDP), depending on different scenarios. The results are summarised in Table 1.5.

Present surpluses of agricultural commodities and the agricultural price support systems in the US and European Union (EU) depress world agricultural commodity prices, so the welfare effects of changes in yields are not so great as the physical effects due to reductions in taxpayer and consumer costs in those areas, and due to increases to producer incentives in regions outside those areas. Other studies on the economic effects of climate change (as reviewed by Fankhauser, 1992) tend towards the negative estimates presented above.

The spatial location of agricultural commodity production is also determined by the world commodity trading system. The effects of changes in relative prices have been considered in economic models of international commodity trade. Pressure to reform the present regimes, where agriculture in the EU, US and

Table 1.5   Economic costs and benefits of changes in world agriculture due to climate change

| Region | Net welfare (% of GDP 1986) | |
	Climate change scenario A	Climate change scenario B
US	0.005	−0.31
EC	−0.019	−0.40
World	0.01	−0.47

Source: Kane et al. (1992).

Notes: Scenario A yield changes include − 10 to − 15% wheat and corn yields in North America and EC but positive (10 to 15%) in Australia and the former USSR. Scenario B yield changes include − 20 to − 40% wheat or corn yields in US and EC and negative (− 10 to − 20%) in USSR and Australia.

elsewhere is supported through intervention pricing and where export subsidies reduce world market prices, comes from both the GATT process for trade liberalisation and from within the blocs for the reduction of the budgetary and environmental costs of the present system.

These external political and economic factors have been incorporated in other global models of the world food system, such as that of Rosenzweig et al. (1993) (see also Rosenzweig and Parry, 1994). Intuitively, dynamic economic adjustments to changing climate can compensate for marginal changes, but in the Rosenzweig et al. analysis the predictions of larger climate change from some climate models lead to large negative impacts on the incidence of food poverty and other indicators of stability of the food system. The socio-economic variables of population growth, economic growth as traditionally measured, and trade liberalisation are all central to future states of the world, but changes in them seem to be amplified by increased variability and stress in the climate and agriculture systems.

Climate change commitments

The actual course of future greenhouse gas emissions will be determined by economic and population growth and technological change, as well as by the implementation process of the Climate Change Convention. Wigley and Raper (1992) estimate the warming and sea level changes associated with low-emission and high-emission scenarios to arrive at their range of 1.6–3.0 °C and global mean sea level rise of 35–54 cm by 2100. To isolate the impacts of particular policy measures, Warrick (1993) reports analysis of how, *ceteris paribus*, emission reduction of just the industrialised countries, and of halting deforestation would change future climate and sea level impacts. The results are shown in Figure 1.2.

(1) If all countries continued with a 'business as usual' growth in greenhouse gas emissions, the best projected results at present show an increase in global

20

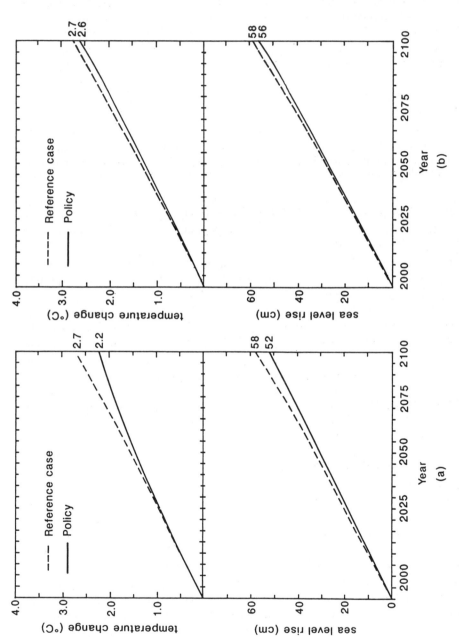

Figure 1.2 The effects of (a) a strong OECD emission control policy and (b) of halting deforestation on global mean temperature (upper panel) and sea level rise (lower panel). Reproduced by permission of the Institute of British Geographers from Warrick (1993)

mean temperature in the order of 2.7 °C and a mean sea level rise of 58 cm by 2100 (Reference line in Figure 1.2).

(2) If the countries of the Organisation for Cooperation and Development (OECD)—primarily high energy using industrialised countries, including the UK—reduced their emissions drastically with a long term aim of making fossil fuel technologies virtually redundant, the global mean temperature would still rise by 2.2 °C and mean sea level by 52 cm by 2100 (Figure 1.2(a))

(3) If deforestation were halted by 2010 (assumed 1.6 bt C year^{-1} present emissions following the IPCC) and reforestation is subsequently pursued enhancing the carbon sink by 1.6 bt C year^{-1}, the resulting temperature and sea level change would be reduced from 2.7 °C to 2.6 °C and from 58 cm to 56 cm by 2100 (Figure 1.2(b)).

The Climate Change Convention is rather vague regarding emission reduction targets, but does suggest that the stabilisation of industrialized nation emissions at 1990 levels by the year 2000 would represent a *starting point* (or, at least, a gesture of good faith towards the developing nations). However, as demonstrated above, this target would hardly meet the stated objective of the Convention of 'avoiding dangerous anthropogenic interference with the climate system', as (a) present concentrations mean that some global warming is already committed to; (b) meeting this target would permit further warming; and (c) even with aggressive emission reduction policies, global warming of some degree is certain to occur.

1.4 TRENDS IN LAND USE CHANGE

Major emissions of greenhouse gases from the land use sector are associated with *changes* in land use. The land use causes of global warming are related in the main to land use change. As illustrated in Figure 1.1 of the global carbon cycle, large reservoirs of carbon exist undisturbed in soil and biomass. The fluxes associated with normal growth and regeneration are small compared to those when land use is intensified or when land use has been changed. The observed trend of accelerating land use change in recent centuries has primarily been driven by frontier expansion and population growth (Richards, 1990).

Table 1.6 shows land use trends from 1700 to 1980, based on various databases and models summarised by Richards (1990). The single most important trend is the reduction in the world's forests and woodlands which are now 81 percent of their 1700 total. The expansion of the cropped land area is almost the same magnitude, though the areas are not coextensive. Table 1.6 also provides evidence that the rate of change has increased dramatically this century, with the absolute changes in world cropland being greater in the period 1950–1980 than in the 250 years before then. Although tropical deforestation receives the greatest attention at present, particularly for its greenhouse gas implications, evidence suggests that over the timeframe of a few centuries, temperate and boreal

Table 1.6 Global land use change 1700–1980

Regions	Vegetation types	Area (million ha)					% changes from				
		1700	1850	1920	1950	1980	1700 to 1850	1850 to 1920	1920 to 1950	1950 to 1980	1700 to 1980
Africa	Forests and woodlands	1396	1370	1302	1206	1088	-1.9	-5.0	-7.4	-9.8	-22.1
	Grassland and pasture	2175	2180	2203	2227	2218	0.2	1.1	1.1	-0.4	2.0
	Croplands	64	84	131	202	329	31.3	56.0	54.2	62.9	414.1
North America	Forests and woodlands	1016	971	944	939	942	-4.4	-2.8	-0.5	0.3	-7.3
	Grassland and pasture	915	914	811	789	790	-0.1	-11.3	-2.7	0.1	-13.7
	Croplands	3	50	179	206	203	1566.7	258.0	15.1	-1.5	666.7
Latin America	Forests and woodlands	1445	1420	1369	1273	1151	-1.7	-3.6	-7.0	-9.6	-20.3
	Grassland and pasture	608	621	646	700	767	2.1	4.0	8.4	9.6	26.2
	Croplands	7	18	45	87	142	157.1	150.0	93.3	63.2	1928.6
China	Forests and woodlands	135	96	79	69	58	-28.9	-17.7	-12.7	-15.9	-57.0
	Grassland and pasture	951	944	941	938	923	-0.7	-0.3	-0.3	-1.6	-2.9
	Croplands	29	75	95	108	134	158.6	26.7	13.7	24.1	362.1
Other Asia	Forests and woodlands	588	569	536	493	415	-3.2	-5.8	-8.0	-15.8	-29.4
	Grassland and pasture	314	312	304	295	279	-0.6	-2.6	-3.0	-5.4	-11.1
	Croplands	57	78	119	171	265	36.8	52.6	43.7	55.0	364.9
Europe	Forests and woodlands	230	205	200	199	212	-10.9	-2.4	-0.5	6.5	-7.8
	Grassland and pasture	190	150	139	136	138	-21.1	-7.3	-2.2	1.5	-27.4
	Croplands	67	132	147	152	137	97.0	11.4	3.4	-9.9	104.5

Former USSR	Forests and woodlands	1138	1067	987	952	941	−6.2	−7.5	−3.5	−1.2	−17.3
	Grassland and pasture	1068	1078	1074	1070	1065	0.9	−0.4	−0.4	−0.5	−0.3
	Croplands	33	94	178	216	233	184.8	89.4	21.3	7.9	606.1
Pacific countries	Forests and woodlands	267	267	261	258	246	0.0	−2.2	−1.1	−4.7	−7.9
	Grassland and pasture	639	638	630	625	608	−0.2	−1.3	−0.8	−2.7	−4.9
	Croplands	5	6	19	28	58	20.0	216.7	47.4	107.1	1060.0
TOTAL	Forests and woodlands	6215	5965	5678	5389	5053	−4.0	−4.8	−5.1	−6.2	−18.7
	Grassland and pasture	6860	6837	6748	6780	6788	−0.3	−1.3	0.5	0.1	−1.0
	Croplands	265	537	913	1170	1501	102.6	70.0	28.1	28.3	466.4

Source: Adapted from Richards (1990, p.164)

deforestation is by far the most significant in areal terms. In another global land use assessment, Matthews (1983) estimates that from the pre-industrial era to the present, tropical forests have declined in area by approximately 3.9 percent where temperate and boreal forests have lost close to 20 percent of their area.

A major driving force behind these changes in the last 300 years is undoubtedly the increased productivity of labour in exploiting land through the application of capital and new technologies. The agricultural revolution and the opening of frontier land in particular explain the conversion of land from natural to agricultural use. Although the causality between population growth, technological innovation and land use change is unclear, population growth is certainly a driving force behind increasing productive use of land (Boserup, 1965) and, by definition, of the intensity of land use. Over the same period of land use change in Table 1.6 (1700–1980), global human population increased more than sevenfold, from 0.6 billion to 4.43 billion.

The impact of present land use change trends on greenhouse gas fluxes is summarised in Table 1.7. The specific trends and their impacts are dealt with in detail in the following chapters. But it is clear that although there are some important trade-offs between the gases, as highlighted in Chapter 4 with regard to natural wetlands, changes in land use have a detrimental effect by causing net emissions of the greenhouse gases.

The FAO (1991) have estimated, based mainly on soil characteristics, the potentially cultivable land in those areas which have the greatest population growth rates. As summarised in Table 1.8, the Asian continent utilises possibly 90 percent of its cultivable land, given present technology and demand for agricultural outputs, so many of the land use changes in that continent in future are likely to be changes in the intensity of use. For rice cultivation for example, this means increasing the number of crops per year as well as adopting higher yielding varieties and importing more nutrients into the system via fertilisers. The implications for greenhouse gas emissions are in the main to cause increases in methane emissions (Table 1.7) and are examined in detail in Chapter 4.

The scope for cultivating more land in Africa and Latin America would seem to

Table 1.7   Greenhouse gas flux impacts of current land use trends

Current land use trends	CO_2	CH_4	N_2O	H_2O	Albedo
Increasing deforestation	+	+ (with burning) 0 (with clear-cutting)	+	−	+
Increasing area of paddy rice	0	+	+/−	0	0
Increasing N-fertiliser consumption	0	0	+	0	0
Increasing desertification	+	0	−	−	+

Source: Adapted from Bouwman (1990c, p.10). na = not available
Note: + = increased emission; − = decreased emission.

be greater than for Asia, based on the FAO physical assessment. The data in Table 1.8 point to a potentially misleading dichotomy between a land scarce Asia and a labour and resource scarce Africa and Latin America, where future expansion of agricultural production in Africa and Latin America will be as a result of expansion of the cultivated area and in Asia will be due to an increase in the intensity of land use. This dichotomy is misleading because of the skewed nature of land ownership and control, meaning that many decision-makers (farmers and land users) are actually land scarce in the African and Latin American continents *and* labour scarce in Asian countries. In reality the mechanisms for the expansion of agricultural production will be a combination of both intensification and expansion of the agricultural frontier. The extent of the expansion has been estimated in Cole et al. (1993) to be of the order of 15 million ha per year, given projections of global population increase to 2025. The implications for greenhouse gas fluxes and the opportunities for abatement are discussed in the various chapters of the book. They will incorporate both agricultural and forest management practices and the implications of the expansion of the agricultural frontier as a cause of deforestation.

1.5 SOCIAL SCIENCE ANALYSIS OF GLOBAL ENVIRONMENTAL CHANGE

The proximate causes of land use change and environmental degradation take place through the private decisions of individuals. The impact of land use change could be said to represent a problem only when the decisions of sufficient numbers of users or owners coincide. In other words land use change is a cumulative phenomenon. However, global warming involves change to a global system with impacts at the global scale, and is therefore different in character to

Table 1.8   Present use of agricultural land resources by region

Region	% of world population	Potentially cultivable area (million ha)	% of total land area	Presently cultivated area (million ha)	Presently cultivated as % of potentially cultivable area	% irrigated of present cultivated area
South-west Asia	3	48	7	274	92	16
South-east Asia	28	297	33	274	92	24
Central Asia	22	127	11	112	89	44
South America	6	819	46	124	15	6
Central America	3	75	27	36	48	18
Africa	10	789	27	167	21	4

Source: FAO (1991).

changes brought about by cumulated local impacts. Turner et al. (1990) explore this issue and define two types of global environmental change: *systemic* and *cumulative* change. Systemic change refers to change to the operation or functioning of a global system, whereas cumulative change is global only when accumulated localised change affects the major parts of the globe. Thus release of carbon to the atmosphere causes systemic change, though the point is made that global scale activity is not necessary for this disruption of the function of the atmospheric system to take place. Thus approximately 56 percent of global emissions of greenhouse gases have their sources in seven countries (World Resources Institute, 1992).

Systemic global change is in a sense analogous to Hardin's Tragedy of the Commons (Hardin, 1968). Every human 'grazes' the atmospheric commons through their actions, as the atmosphere has now gone beyond its assimilative capacity and each action which adds to the atmospheric concentrations of the greenhouse gases further enhances the greenhouse effect. As Hardin's paper was written in the context of increasing global human population and argued that each child added to the impact on the commons, some of the analogies of the atmosphere as a global open access commons do bear up. However, Hardin's ideas were disputed and discredited for numerous reasons (O'Riordan and Turner, 1983), including the social and moral norms which exist in commons situations; and the non-alignment of those exploiting most commons with those who bear the impacts. These issues are explained in economic terms by the concept of public and private externalities.

Baumol and Oates' (1988) definition of an externality requires that those whose actions cause the impacts do not pay in compensation for this impact an amount equal in value to the resulting cost to others. Interaction with the global geochemical cycles which cause the increases in atmospheric concentrations of the greenhouse gases clearly exhibits externality characteristics. Such externalities would be fully internalised if the present international actions under the Climate Change Convention compensated *all* the impacted parties *fully*, but this is clearly unattainable for a number of reasons. The first is the intergenerational nature of the climate change problem, and the second is the complex interaction of private good externalities and public good externalities which make up the climate change issue.

The greenhouse gases are ambient pollutants in that they mix evenly throughout the atmosphere. The location of the emission of one unit of any of the gases is unimportant in terms of the greenhouse effect, though *when* the emission occurs (now rather than later) obviously is important. The greenhouse gases are therefore stock pollutants, with the increases in the stocks having a positive relationship with their future impacts. The impacts of the emissions (primarily global climate change and sea level rise), however, are not evenly distributed over the globe, and their impact on human welfare is likely to result in even greater mal-distribution. So the global climate change problem has some public and some private implications, as set out in the Table 1.9.

Table 1.9   External impacts of global climate change

Private impacts	Public impacts
Impacts on resources, land, agriculture and forests	Uncertainty and societal risks
	Perceptions of a tainted earth
Risks to individual health	Impacts on other species
Risk of consequences of extreme events and insurance against them	

The impacts of climate change which affect private decision-makers are those which impact on present and future income and wealth and also on private risks of future impacts such as health. These impacts are cumulative and are those identified in Table 1.9. They include impacts on agricultural production, changes in the use of energy and the cost of insurance. The 'public' impacts of global warming by contrast are those to which no-one is immune. These are for example the impacts on society (rather than on the individual) of drought or flooding; the impacts on non-human species which are generally only accounted for in analysis as their secondary impact on human welfare; and societal perceptions that the whole of the natural system is nowhere unpolluted. This latter 'loss of wilderness' impact of global warming has been eloquently described by Bill McKibben in his *The End of Nature*:

> If you climbed some remote mountain in 1960 and sealed up a bottle of air at its peak, and then did the same thing this year, the two samples would be substantially different. The air around us, even where it is clean and smells like spring and is filled with birds, is *different*, significantly changed.... The idea of Nature will not survive the new global pollution. We have deprived nature of its independence, and that is fatal to its meaning. Nature's independence *is* its meaning; without it there is nothing but us.
>
> (McKibben, 1990, pp. 17, 54)

The distinction between public and private impacts of global environmental change and of cumulative and systemic change are vital in determining strategies for adapting to and, in the context of this book, for mitigating the processes which lead to global warming. In economic terms, greenhouse gas abatement will be carried out by *individuals* reacting to the prevailing economic and social framework and policies designed to reduce emissions. Pressure to create these incentives will come about because the cumulative private and public impacts of global warming will be perceived by individuals and by countries.

Economics and policy analysis

Given the preceding discussion of the nature of the greenhouse problem and human interaction with the biosphere as a cause, this section illustrates how this problem could be reduced or adapted to, and the implications for society of these proposed solutions. The approaches taken in this book are those of welfare economics and of political economy. Neo-classical welfare economics provides a framework which is necessarily anthropocentric and which takes individual human preferences as its focus. Actions concerning the human use of land and the natural environment are viewed from the perspective of their impact on the present and future welfare of those using the resource and those affected by its use. The concept of externality has already been introduced and defined as an impact on those not owning or controlling the land or other resource. If greenhouse gas emissions reduction is to take place, the externalities created by the impacts of climate change on future generations have to be taken into account. So although it may be beneficial in terms of greenhouse gas fluxes to afforest an area half the size of the United States to offset current CO_2 emissions (Sedjo and Solomon, 1989) this should only be undertaken if this was a cost-effective way of reducing net emissions, and if those people who lose income or welfare by carrying out such a policy are identified and possibly compensated.

Sustainable development

In all proposed actions, the economic costs to society should be weighed against the economic benefits to society. However, the concept of sustainable development suggests that simply taking the economic costs and benefits into account when considering responses to the greenhouse effect is not sufficient. Sustainable development is the central theme of both the Climate Change Convention and the other conventions signed at the Earth Summit in 1992. The sustainability criteria which any development action should meet (as summarised by Brown et al., 1993) include:

(a) intergenerational equity: as implied in the Brundtland Commission's definition;
(b) ecosystem resilience: this stresses the interdependence of natural systems and the processes which bring about global environmental change;
(c) equity in opportunity and human development: this is not only a desirable goal in itself, but also necessary to safeguard intergenerational equity in resource use. This condition stresses the role of sustainable livelihoods, and the importance of human rights and empowerment in bringing about sustainable development.

These criteria are wider than the often quoted definition of sustainable development from the Brundtland Commission (World Commission on Environment and Development, 1987) that development is sustainable if it '... meets the needs of the present without compromising the ability of future generations to meet their own needs'. Essentially the approach in this book accepts that feasibility, cost and equity (in both inter- and intra-generational senses) are the criteria by which abatement should occur.

The principles of sustainable development outlined above directly impact on how decisions concerning abatement should be taken. As argued by Howarth and Monahan (1993) among others, applying the principles of sustainable development implies that the optimal level of abatement is not sufficient to meet the sustainability criteria. The application of equity rules as discussed above is critical both to analysing who should abate, both in the land use and non-land use sectors; and who should compensate those most likely to be affected by the consequences of climate change.

Examples of the social benefits of greenhouse gas reduction and the least-cost policy measures in the land use sector are given in Part III. As has been discussed here, these criteria fulfil only one aspect of sustainable development. As pointed out by Buttel (1993), it is important to recognise the human and policy dimension behind the characterisation of land and vegetation as simply part of global geochemical cycles. The debate on 'sustainable' agriculture in the United States has encouraged spurious 'environmental' factors to be introduced. The implications of this for sustainable development in its widest sense are insidious:

> What then will be the future of rural America if it becomes defined in strong symbolic terms as forest sites or prospective forest acreage needed to curb the greenhouse effect ...? Will we, in other words, witness a further erosion of commitment to improving the livelihoods of the rural poor and to rural development?
> (Buttel, 1993, p.24)

Thus although policy can be undertaken to deal with environmental problems directly, and the environment can be taken into account in government action with other objectives, the aims of sustainability require the incorporation of both environmental constraints and objectives and those of social and inter-generational equity.

Cost

As outlined above, the emissions of greenhouse gases can be analysed as an externality problem in that the full social cost of emissions is not accounted for. The processes or activities giving rise to emissions of greenhouse gases are exacerbated by a series of failures of markets to reflect the full social cost of emissions, and of failure of interventions in markets by government through regulation, taxation or subsidies to send the 'correct' signals to individual decision-makers. The removal of these market and intervention failures can mean that the cost of reducing emissions at the margin is very low or even negative. The identification of these opportunities across all sectors is important, and if further abatement is to take place, the least-cost options should be undertaken first. Welfare economics uses the concept of equating marginal costs and benefits to determine the optimal rate at which action should be taken to maximise the aggregated well being of 'society'.

Standard economic analysis for the global case can be illustrated by the marginal abatement cost and marginal social benefit diagram (Figure 1.3), which shows the optimal level of abatement of greenhouse gas emissions (taking the costs of reducing emissions and the benefits of avoiding global warming) at the global level. The optimal level of abatement of the emissions of greenhouse gases is then found at X, which equates the marginal cost of abating these emissions (MC) with the marginal benefit (MB) of avoiding climate change impacts. Moving from the point A, which can be thought of as a business as usual scenario, provides large marginal benefits with initially low marginal cost. This may be due to large external as well as direct benefits—the benefits of reducing emissions of CFCs, for example, are greater than just its greenhouse potential. The negative section of the MC curve (AB) in Figure 1.3 indicates the existence of no-regret measures such as energy conservation which are internally cost-effective (Lovins and Lovins, 1992).

Generally, the optimal level of policy action against the greenhouse effect depends on the costs of the future impacts, and on the costs of the policies considered. In this section we distinguish between *first best* global solutions to global climatic change and those of a partial nature, only applicable at the regional scale. For the global problem, the question is one of determining the

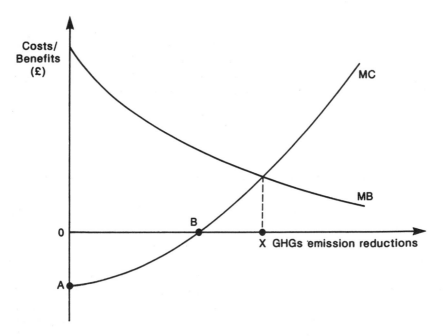

Figure 1.3   Economic costs and benefits of greenhouse gas abatement. MC = marginal costs of abatement; MB = marginal benefits to society of averted climate change damage; 0X = economically efficient level of emissions reduction

optimal level of greenhouse gas abatement. In neo-classical economic terms, the marginal abatement cost should be equated with the marginal damage function (Figure 1.3) or, adopting a precautionary approach, thresholds should be set above which the risk (or cost) should be deemed undesirable. Various analyses have attempted to estimate this global optimal abatement (Nordhaus, 1991, 1992; Cline, 1992; Peck and Teisberg, 1992). The results show that the optimal level of abatement may range from 10 to 30 percent of present emissions, though the studies are sensitive both to the estimates of the narrow economic definition of the damage function and to the discount rate used. These issues are further discussed in Chapter 6. At the local level, the abatement of greenhouse gas emissions is not relevant to mitigation strategies. Changes in local emissions are too small to have a noticeable impact on climate and the decision is thus only concerned with protection and adaptation costs to exogenous likely, or even already committed to, scenarios of climatic impacts or accelerated sea level rise.

The implementation of prescriptive measures based on estimates of abatement costs will be dependent on whether precautionary views or strict efficiency views are adopted, and on the presence of uncertainty in the MC or MB curves (Baumol and Oates, 1988). If the absolute slope of the MB curve is greater than that of the MC curve and uncertainty exists in these abatement costs, then a policy with absolute limits on emissions (such as a tradeable permits system) is more desirable than a tax-based system. Alternatively, where the MC curve is more steep than the MB curve over the relevant range, and uncertainty exists, then a tax system is desirable as large welfare losses are involved in instigating overstringent abatement. In other words, with a tax, the absolute amount of emissions is unpredictable over time and the tax may need to be adjusted to meet particular targets. With a quota system, the absolute limit is predetermined based on information concerning possible catastrophic events or for political feasibility.

At the international level, there would be obvious institutional differences in the way these systems operate. From the experience of other market and regulatory mechanisms for pollution control, it would seem that influence-seeking institutions prefer direct regulation to market or quasi-market based systems, given their more direct control over setting and enforcing standards (Hahn, 1989). In the greenhouse arena, no market-based solutions have seriously been considered (though their design and potential impacts have been analysed by many on the conceptual level) (Bertram, 1992; UNCTAD, 1992; Grubb, 1993), with the exception of using fiscal measures within one country or region, such as the proposed EU carbon tax (Carraro and Siniscalco, 1993).

In conclusion, land use strategies to abate greenhouse gas emissions are only attractive when the costs of other actions, in the energy sectors or in bioengineering solutions for example, are greater than the net costs of afforestation or reduction of agricultural emissions. The role of uncertainty is crucial in determining the optimal strategy, given the cost and efficiency criterion.

Technical feasibility

The feasibility of reducing emissions or enhancing sinks has been prominent in many studies of 'solutions' to reduce greenhouse gas emissions. As Figure 1.1 illustrates, there are many interactions between the terrestrial and oceanic parts of the biosphere and both interact with the atmosphere. Interventions at most of these points have been suggested. Oceanic photosynthesis for example may be removing 35 bt C year^{-1} although a large proportion is returned to the atmosphere through decomposition of marine organisms. Strategies for intervention in this cycle concentrate on enhancing the rate of maritime carbon fixation by adding iron to the ocean surfaces to stimulate phytoplankton blooms. It has been estimated that 2–3 bt C year^{-1} could feasibly be sequestered by this process at a cost for the iron and shipping of up to US \$50 per tonne carbon. However, there are possible adverse effects to enhancing oceanic blooms. Presently increasing algal blooms have direct economic implications for fisheries and aquaculture through deoxygenation of fisheries, and potentially on human health through shellfish poisoning and through potential links to cholera (Epstein et al., 1993). Further, the mechanisms by which the carbon would be transported to the deep ocean if deliberate seeding took place have been seriously questioned (see Grübler et al., 1993). Other schemes such as changing the reflective properties of the earth's surface, and collecting and disposing of CO_2 from flue gas streams also seem far-fetched and difficult to assess in terms of net effect.

The major areas for serious consideration are in changing the use of energy resources and the use of land resources. Some energy-related interventions, such as increasing non-emitting technologies such as electricity generation from water also have implications for land use (Rosenberg and Scott, 1994). The primary interventions in the land use sector as already described are in stopping present deforestation, in afforestation and in reduction of methane emissions from agricultural sources such as livestock and paddy rice cultivation.

Equity

The cost-effective solution to abating greenhouse gases, as described above, is to carry out negative and least-cost solutions first, up to the point where the extra cost of doing so outweighs the benefits of doing so. However, whichever incentives or regulations are used to reduce emissions, the distribution of the costs will not be evenly spread, raising issues of *who* should pay. Further, at the international level, equity issues are pertinent in the 'common but differentiated responsibility' clause of the Climate Change Convention. Any rule or system which can be devised to incorporate equity (current emissions, historical emissions, emissions per capita and other criteria) will address where the responsibility for global warming lies. If a present-day 'polluter pays principle' is adopted this ignores the role of historic emissions in causing the present-day concentrations of the greenhouse gases, and is therefore regarded as being unfair to presently non-industrialised countries (see for example Rose, 1992).

1.6 THE INTERNATIONAL CONTEXT AND THE CLIMATE CHANGE CONVENTION

The clearest recognition of the international concern over climate change comes in the form of the United Nations Framework Convention on Climate Change (the Climate Change Convention), signed in 1992 by 155 states and subsequently by a dozen more. The UN Conference on Environmental and Development in Rio de Janeiro was the culmination of a process resulting in the signing of a Biodiversity Convention, the Rio Declaration, Agenda 21 and a set of Forest Principles as well as the Framework Convention on Climate Change. All these statements and conventions recognise the context of human interaction with the biosphere and environmental change proceeding more rapidly at present than at any time in the past. The impetus for the Climate Change Convention was the growing scientific consensus, loosely defined, as to the nature of the greenhouse effect and its potential impacts. Additionally, political impetus came from extreme climate events, principally the US drought of the late 1980s (Grubb et al., 1993).

Climate Change Convention

The Climate Change Convention is the single most definitive statement of intent of the majority of countries in the world to minimise their impact on the climate system by reducing their emissions of greenhouse gas emissions and to assist the most vulnerable areas to adapt to climate change. Although many countries now interpret the targets agreed to in the Convention as their ultimate and only goals, the process of the Convention has been envisaged as one which would lead to greater action over time as greater scientific understanding is reached. In many ways this reflects experience with other international treaties such as the Montreal Protocol on controlling substances which deplete stratospheric ozone. However, the greenhouse gas issue is significantly different to the Montreal Protocol process in that greenhouse gas emissions have diverse sources which are directly coupled to economic activity, primarily the use of fossil fuels. The scope for radically altering the structure of economic activity worldwide is somewhat limited and may not simply be brought about by international treaty.

The ultimate objective of the Climate Change Convention is:

> the stabilisation of greenhouse gas concentrations in the atmosphere at a level that would prevent dangerous anthropogenic interference with the climate system. Such a level should be achieved within a time frame sufficient to allow ecosystems to adapt naturally to climate change, to ensure that food production is not threatened and to enable economic development to proceed in a sustainable manner.
>
> (Climate Change Convention, Article 2)

This objective raises more issues than it resolves in that it implicitly accepts that some climate change is inevitable, given the present commitments; that the adaptation of natural ecosystems is important even though this is one of the greatest sources of uncertainty of impact; and that sustainable development is an ultimate social goal. The objective begs the question as to whether sustainable development and meeting the objectives of the Convention are compatible with each other. This is raised in later principles of the Convention when one of the paragraphs seemingly acknowledges certain trade-offs:

> The Parties have a right to, and should promote, sustainable development. Policies and measures to protect the climate system against human-induced change should be appropriate for the specific conditions of each Party and should be integrated with national development programmes, taking into account that economic development is essential for adopting measures to address climate change.
> (Climate Change Convention, Article 3, para. 4)

The signatory Parties have agreed in principle to move towards the stabilisation of all greenhouse gas emissions at 1990 levels by 2000. However, as with the other agreements signed at Rio, 'sustainable economic development' is an overriding priority and allows developing countries to increase their emissions in the short run. Each of the signatory parties to the Convention has agreed to a number of specific mechanisms to achieve the stabilisation. These include the formulation of national inventories of all greenhouse gas emissions and sinks and the presentation of these, along with national plans to the Conference of the Parties set up under the Convention, and to promote technology transfer to assist in both abatement of emissions and adaptation to climate change. As developing countries are almost certain to increase their emissions during periods of economic development, developed country parties would have to reduce their emissions to give them this 'space' (Hayes and Smith, 1993a). Compliance can be brought about 'individually or jointly' between countries or regions. This realises potential for international cost-effective abatement of emissions, but raises questions of enforcement and of equity. The scope for joint implementation agreements would seem to be limited due to the transaction costs involved in setting up cost-effective abatement arrangements. Moreover, large-scale adoption of these arrangements could lead to less emission reduction if double-counting between countries occurred.

Despite the targets and apparent commitments of the Convention, it is simply a framework around which the process should be taken forward. The Convention is now a binding international treaty having been ratified by 50 signatory parties. However, national country commitments to the Convention are often overshadowed by the economic and political circumstances, such as the impact on employment, the public perception of instruments such as taxes, and the global economic recession taking place during the signing of the Climate Change Convention and the years beyond. For example, the UK Government is committed to reducing its projected CO_2 emissions to 1990 levels by the year 2000, and has set out a number of measures which will achieve this (see for example Brown and Maddison (1993) and the discussion in Chapter 2). This task is in effect easier to achieve by the present slowdown in the UK economy and the switching of fuel sources for electricity generation from coal to gas as a result of the restructuring of the electricity generation industry. However, the UK has opposed the introduction of an EU carbon energy tax, which other EU member states feel is necessary to achieve the targets of the Climate Change Convention, primarily because it can meet its obligations to the Convention without the imposition of the tax. So although the Climate Change Convention is the single most important international manifestation of concern about the climate change threat, the action of national signatory parties is more likely to be dominated by considerations of domestic acceptability of abatement strategies and by their perceptions of international co-operation within the Climate Change Convention framework.

1.7 CONCLUSION

Land use change is a continuous, evolving process, and is the single most important manifestation of human interaction with the biosphere. The scale and rate of change of land use is greater now than at any time in history due to rapid technological change and to population growth. Land use practices and land use change play an important contributory role in the human impact on the atmosphere and ultimately on the whole biosphere.

The rest of this book takes up this theme. First, Part II sets out the mechanisms of how land and land use activities impact on the global cycles of the greenhouse gases. These chapters show the areas of uncertainty as well as illustrating methodologies for eliciting fluxes, given the current state of knowledge.

Chapter 2 discusses how land use change data can be used to estimate aggregate net fluxes of greenhouse gases, based on simple models of vegetation growth and the interaction of soil and biomass. Aggregate fluxes are estimated for the UK, showing that carbon accumulates on that portion of land remaining in the same use over time, but that disturbance in general causes net emission.

Chapter 3 examines the CO_2 and other fluxes associated with tropical forests in particular. This is the single most important land use change in terms of the greenhouse effect, yet large degrees of uncertainty still exist over areas of forest loss and the associated greenhouse gas fluxes. Estimation of the fluxes associated with forest loss in the African continent in the last decade is presented, utilising the methodology set out in Chapter 2 and the latest FAO global forest assessment data.

Chapter 4 further examines greenhouse gas fluxes, principally CH_4, associated with natural and artificially created wetlands. Both are important sources of CH_4 emissions, though artificially created wetlands areas offer more scope for management intervention. A new aggregate estimate of global emission of CH_4 from rice and coarse fibre production is presented, which is higher than previous global estimates due to the influence of a high estimate for China, based on a best fit of country emission rates.

Part III adopts a social science-based analysis of the human factors involved in these actions. The causes of deforestation are explored in Chapter 5; the scale of economic activity in frontier agriculture is a fundamental cause but this is a complex set of actions including logging and clearing for agriculture, exacerbated by policy failures. Policy failures leading to non-sustainable resource use are caused by the design of institutions and public policy pandering to interest groups.

Chapter 6 estimates the *benefits* of reduction of CH_4 emissions in the livestock sector of the UK. As agricultural price support is a significant proportion of the income from rearing livestock, the real cost of reducing support is not as high as it would appear. If the benefits of reducing CH_4 emissions are valued at the monetary value of global warming damage avoided, agricultural policy reform can be further justified.

Chapter 7 estimates the cost of influencing net emissions through afforestation. The costs of afforestation vary greatly across the world. A further UK example shows that a critical parameter is the availability and cost of land. Afforesting all land currently diverted from agricultural production in the UK would only reduce overall emissions by a small fraction (less than 1 percent). Large-scale afforestation has many other economic, social and environmental consequences, some of which are clearly at odds with offsetting greenhouse gas emissions.

Chapter 8 highlights international mechanisms for greenhouse gas reduction. It focuses on joint implementation under which forest projects are currently being implemented to offset the emissions of one country by sink enhancement in another. It concludes that although theoretically beneficial, joint implementation should not form part of a global strategy because of the impacts of afforestation and the need to ensure that profligate emitting countries enact policies which reduce domestic emissions. We conclude in Chapter 9 with a summary of options for emission reduction and conclusions on the nature of the global warming problem.

PART II

THE LAND USE CAUSES OF GLOBAL WARMING

CHAPTER 2

Assessing Greenhouse Gas Fluxes from the Terrestrial Biota

2.1 INTRODUCTION

The nature of land use is continually changing over time by either natural or human-induced means. To assess the implications of this for fluxes of the greenhouse gases, data on temporal changes are required. As discussed in Chapter 1, the principal land use changes globally are generally towards agricultural uses from semi-natural or natural types. The impact of these changes on the carbon cycle is on the storage of above-ground biomass and on soils.

In tropical countries a very significant proportion of the fixed carbon is in the form of the above-ground biomass of forests (Brown and Lugo, 1992; see Chapter 3). In temperate zones, a greater proportion of the fixed carbon in the land use sector is held in soil organic matter than in the standing biomass. Changes in the soil carbon pool may have an effect in temperate areas comparable to, or greater than, the retention of carbon in living tissue above ground. Although biological processes fix carbon from the atmosphere, it is common for much of this to be returned after one growth cycle, either through digestion or waste. Only if carbon remains in a fixed form over long periods can biological processes make a contribution to the accumulation of carbon. Examples of this are peatbogs and in forestry.

In this chapter the difficulties with deriving fluxes from inadequate land use data are discussed. Land use change data sources tend to be either from remote-sensing techniques such as satellite imagery or aerial photography, or from sample field survey. Aggregate yearly carbon fluctuations for Great Britain are estimated. This is based on an aerial survey of Great Britain with two time points in the last 50 years. The results show that conversion of land to arable cultivation leads to medium term reductions in soil and biomass carbon, when the limits of the study are set to count agricultural and forestry products leaving the sector as having been fixed by that sector. Forestry in the period studied has not fixed large amounts of carbon because new planting initially causes losses of previously fixed biomass and soil carbon. The aggregate estimates of carbon sequestration are critically dependent on the boundaries of the study, but the estimates are still relatively minor in relation to total carbon emissions from other sectors of the economy in Great Britain.

2.2 LAND USE CHANGE DATA

Detailed global land use data are notoriously unreliable or non-existent. This can be seen from an inspection of international sources such as the UN, FAO or UNEP studies, or compilations of environmental data such as *World Resources* (WRI, 1992) or *Environmental Data Report* (UNEP, 1991). Although economic data are closely monitored, the further one gets away from measures of wealth and consumption, the less reliable the data. The agricultural sector is of vital importance to the economy of many countries, so agricultural production figures tend to be collected on an annual basis, especially for the purposes of predicting food import requirements or likely export revenue. Land use data, such as the proportion of a country's land in agricultural use, the use of fertiliser, or forest cover, are usually available from agricultural censuses. But such censuses are carried out perhaps every decade in developing countries, and as pointed out by Bilsborrow and Goeres (1994), are often not carried out at all when budget crises occur and resources are scarce.

For non-productive land uses, such as wetlands and virgin forests, the data are even less reliable. These non-disturbed habitats are important in the global cycles of the major greenhouse gases, as well as for their other environmental services. Global estimates of methane emissions from natural wetlands, for example, rely on databases such as the UNESCO vegetation atlas (Matthews and Fung, 1987) with 28 wetland vegetation types, where Aselmann and Crutzen (1989), using different data, defined 45 categories of wetlands (see Chapter 4). Data difficulties with deforestation rates are also numerous. In addition to the definitional problems of forest categories, the technology of satellite imagery of rates of loss of forest cover also poses problems. For Brazilian Amazonia, for example, satellite data derived estimates of deforestation for 1987 range from 1.7 to 8.1 million ha (Grainger, 1993) (see Chapter 3). Indeed this discrepancy fuelled the debate over responsibility for global warming, when the World Resources Institute index of global warming used 1987 deforestation data for Brazil, this year having an abnormally high rate of deforestation (e.g. WRI, 1990; Ahuja, 1992). The problems of satellite data collection, which is the only option for global scale data for detecting land use change, are formidable. Remote sensing, which relies on solar and thermal wavelengths, is weather dependent, though microwave sensors overcome this problem. All systems have difficulty in determining accurate deforestation rates in Amazonia, especially when most forests were being lost through burning and smoke particles obscured the vegetation cover.

For temperate countries, such as the UK, the weather problems for solar and thermal wavelengths mean that for cloudy areas such as the north of Scotland, only one or two cloud-free images are consecutively available over a period of several years (Mather, 1992). Integration of land use data for the UK is presently underway through the co-ordination of field survey, satellite images and other data sources (Bunce et al., 1992). This integration allows detailed analysis of carbon fluxes and other impacts relevant to the study of climate change. This is presently being carried out with the results of the 1990 Countryside Survey which allocates land into 32 land classes (see Parry et al., 1992; Bunce, 1993).

Two factors required for land use data to be of use in analysing biogeochemical cycles are:

(1) time series data, rather than a single cross-sectional survey;
(2) identification of the changes between categories or classes of land use.

The methodology and results analysing land use change in Great Britain presented below use a database compiled from aerial photography. As with data for any region, the criterion of time dependence is necessary, though the results presented are essentially a comparative static analysis. The general principles of identifying land use change and assessing *net* carbon and other fluxes are applied in the examples throughout this book.

2.3 LAND USE CHANGE DYNAMICS

The information is presented in square matrix form, enabling the isolation of the various elements of land use change, i.e. the land retained in the same use, and land changing into and out of the particular categories. If the land use transition matrix is A_{LU} then:

$$A_{LU} = \begin{pmatrix} a_{11} & a_{12} & \cdots & a_{1n} \\ a_{21} & a_{22} & \cdots & a_{2n} \\ \cdots & \cdots & \cdots & \cdots \\ a_{m1} & a_{m2} & \cdots & a_{mn} \end{pmatrix}$$

where

$m = n$,
$i = 1, 2 \ldots n$ land use categories.

The important elements from such a matrix in terms of carbon fluxes are the leading diagonal elements which represent land in the same category throughout the transitional period. The fluxes associated with soil and biomass carbon for these land uses are equilibrium accumulations. For example, in the forestry categories of land use, trees accumulate carbon over time as the age of the forest lengthens. Similarly for land use categories such as arable, there is no net accumulation of carbon over time in above-ground biomass, though soil organic content usually falls with intensive arable use. These leading diagonal elements can be defined as β_i:

$$\beta_i = a_{ij} \qquad\qquad \forall\, i = j$$

The off-diagonal elements represent changes in land use, and it is in these elements that large fluxes of organic matter occur. Land use changes from the category in the row to the category in the column. The conversion impact factors for changing land use (δ) into each land use category are given by:

$$\delta_i = \sum_{i=1}^{m} a_{ij} \qquad\qquad \forall\, a_{ij} : i \neq j$$

The net effects of land use retained in the same use over the period and of land converting into other categories are seen in the sum of the columns, which is the land in the particular category i at the end of the period (γ). For each category this is given by:

$$\gamma_i = \sum_{i=1}^{m} a_{ij} \qquad\qquad \forall\, i$$

So

$$\gamma_i = \beta_i + \delta_i$$

where γ_i = area in category i at the end of the period.

The relationship between γ, β and δ is represented schematically in Figure 2.1. γ_i, the area in category i at the end of the period, is made up of some land on which

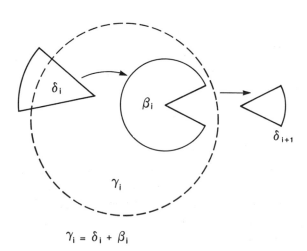

$$\gamma_i = \delta_i + \beta_i$$

Figure 2.1   Categories of land use change (see text)

rapid carbon fluxes are occurring, and other land which is essentially in equilibrium. Given that such a dataset will essentially be a snapshot at different time periods, an assumption required is that the area β remains in the same use, rather than moves out of, then back into the same category in the single time period.

A database of land use change for Great Britain is available. In Table 2.1, a condensed version (from 20 land use categories) of this land use transition matrix is presented. This shows the main land use changes in England, Wales and Scotland between 1947 and 1980 and comes from a number of published sources (Hunting Surveys, 1986; NCC and CCS, 1988; and unpublished data from English Nature). Unfortunately, equivalent data are not available for Northern Ireland, which makes up 5.6 percent of the land area of the United Kingdom, so the results are not on the same basis as the emissions which are based on other human economic activities. The classes presented here separate land use cover into three different types of woodland, two types of semi-natural vegetation, crops, improved grass and a residual category 'other and urban'. The right-hand column represents the distribution of land uses in 1947 whilst the column totals give the distribution for 1980. The lead diagonal elements indicate the percentage of land which was found to be in the same category in both of these years. This is assumed to have remained in the same type of cover throughout the whole period. For forestry stands with average rotations longer than the study period there would be a low probability of land use change occurring more than once, though the probability would increase for non-forestry categories. The off-diagonal

Table 2.1 Aggregated transition land use matrix for Great Britain, 1947–1980 (percentage of total area)

Percent cover in 1947/1980	(a)	(b)	(c)	(d)	(e)	(f)	(g)	(h)	Total 1947
(a) Broadleaved woodland	2.8	0.4	0.2	0.2	0.0	0.3	0.5	0.1	4.5
(b) Coniferous woodland	0.0	0.8	0.1	0.1	0.0	0.0	0.0	0.0	1.1
(c) Mixed woodland	0.0	0.1	0.5	0.2	0.0	0.0	0.0	0.0	1.0
(d) Upland semi-natural vegetation	0.2	2.5	0.4	29.0	0.2	0.7	2.4	0.6	36.0
(e) Lowland semi-natural vegetation	0.1	0.2	0.1	0.1	0.2	0.2	0.2	0.0	1.0
(f) Crops	0.1	0.1	0.0	0.5	0.0	16.6	4.1	1.0	22.3
(g) Improved grassland	0.2	0.2	0.1	1.3	0.0	8.8	17.0	1.5	29.1
(h) Other and urban	0.1	0.0	0.0	0.2	0.0	0.2	0.2	4.4	5.1
Total 1980	3.6	4.2	1.4	31.6	0.4	26.9	24.4	7.6	100.0

Source: England and Wales: Hunting Surveys (1986). Scotland: unpublished land use data supplied by English Nature and Scottish National Heritage. This estimation reproduced by permission of Academic Press Ltd from Adger et al. (1992a).
Notes: Reported standard errors are available for the England and Wales survey only. The error bands are wider for land use categories which occur in larger blocks: woodland (s.e. 7%), semi-natural vegetation (s.e. 10%), farmed land (s.e. 1.5%).

elements indicate movements between classes of land use cover over the three decades to which the table relates.

It can be seen, for example, from row f that crops accounted for 22.3 percent of cover in 1947, of which 16.6 percent was also under crops in 1980. Column f shows that the total amount under crops in 1980 was 26.9 percent—a one-fifth increase in share over the three decades. The sources of this increase in cropped land can be seen by reference to the off-diagonal elements in column f. Thus 0.3 percent of the total surface area was converted from broadleaved woodland to crops, 0.7 percent from upland semi-natural vegetation to crops, and the main source of increase in crops was the transfers from improved grass which account for nearly a third of the 1980 total crops share.

To illustrate the changes for a single land use category over the period, Figure 2.2 shows how the area of the broadleaved forest category, with a significant impact on the carbon cycle, has remained unchanged over 63 percent of its original area, but moved out of the category in 37 percent of its area. The most significant 'destinations' have been into grassland, arable cropland and conversion to coniferous forest. The aggregate net changes between 1947 and 1980 are illustrated graphically in Figure 2.3.

To quantify the carbon fluxes from land use, carbon fluxes associated with soil carbon and standing biomass carbon were modelled (Adger et al., 1992a). The carbon fluxes associated with these land use changes for the 20 categories were estimated as:

$$TF = \sum_{i=1}^{n} \gamma_i \cdot (SF_i + BF_i)$$

or

$$TF = \sum_{i=1}^{n} \beta_i \cdot (SF^{\beta}_i + BF^{\beta}_i) + \sum_{i=1}^{n} \delta_i \cdot (SF^{\delta}_i + BF^{\delta}_i)$$

where

TF = total carbon flux (mt C year^{-1}),
i = 1, 2 ... n land use categories ($n = 20$),
$\Sigma \gamma_i$ = land area of Great Britain (ha),
SF_i = carbon factor for soil under land use category i (t C ha^{-1} year^{-1})
BF_i = carbon factor for standing biomass of land use category i (t C ha^{-1} year^{-1}).

Land retained in the same category over the period, where soil and biomass carbon accumulate towards an equilibrium position, and those areas where land use change has occurred have different profiles of carbon fluctuation and storage which can be summarised as:

(1) *Retained land use* (leading diagonal). Organic matter accumulates towards equilibrium in soil and biomass. The forestry estate moves towards a mature

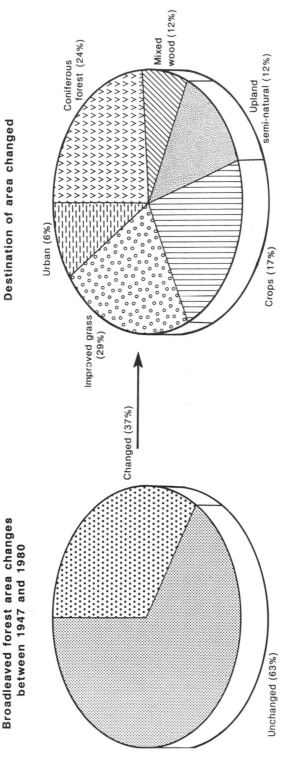

Destination of area changed

Coniferous forest (24%)

Mixed wood (12%)

Upland semi-natural (12%)

Urban (6%)

Crops (17%)

Improved grass (29%)

Changed (37%)

Broadleaved forest area changes between 1947 and 1980

Unchanged (63%)

Figure 2.2 Changes in the area of broadleaved forest in Great Britain, 1947–1980

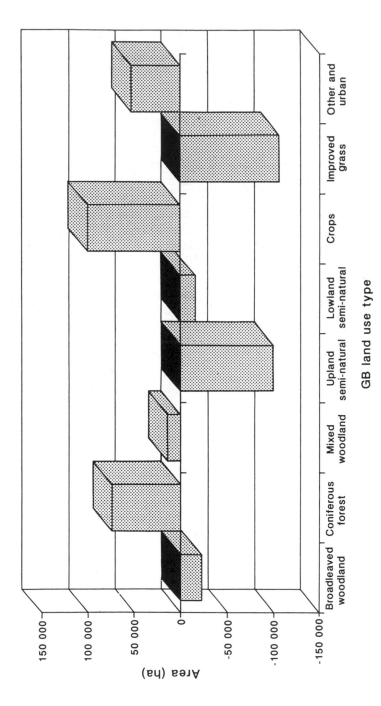

Figure 2.3 Net changes in Great Britain land use, 1947–1980

age distribution, which was modelled using data on the age distribution of the forests. Other land uses reach biomass and soil equilibrium positions more quickly, and for lack of land use change data, are assumed to be in this state for the study period. $BF^{\beta}{}_i$ and $SF^{\beta}{}_i$ for these land uses are therefore estimated on a yearly average change basis.

(2) *Land use change* (off-diagonal elements). A once for all change in biomass carbon occurs some time during the period, and soil and biomass carbon move towards a new equilibrium position. Assumed time profiles of these changes and estimates of the rapid change in soil and biomass organic matter to the new equilibrium are used to calculate average yearly biomass and soil factors (SF_i and BF_i).

Biomass carbon factors (BF_i) are estimated using experimental data on biomass growth, and weights for species and crop mixes in the forestry and non-forestry categories. Similar procedures were undertaken for the soil carbon factors where weightings were estimated for soil types under different land uses (Adger et al., 1992a). Figure 2.4 shows that the effect of land use change on both soil and biomass organic matter is the net effect of both soil and biomass fluxes. This has frequently been overlooked when considering the potential carbon uptake of afforestation schemes (see Chapter 6). When land changes from a semi-natural category such as heath to coniferous forest, there is an instantaneous release of carbon dioxide, as the vegetation is removed and decays or is burned. Assuming that the soil carbon was previously in equilibrium, drawing down of soil organic carbon occurs in the first decade after tree planting, so although coniferous forest accumulates organic matter in its first years after planting, part of this is offset by a reduction in soil organic matter content. In Figure 2.4 net gains only come about after the first decade of planting coniferous forest on heathland. The carbon accumulation then continues at a decreasing rate (in the diagram up to 40 years). Figure 2.4 shows only a single rotation, in that once felling takes place, total carbon accumulation is dependent on the end use of the timber and the amount of waste and loss.

The following sections set out in detail the analysis of the per area fluxes of soil and biomass carbon from land uses and the resulting aggregate fluxes from the land use sector. These fluxes have generally gone unaccounted for in mitigation policy to date in the UK. The fluxes from land use are therefore compared to the aggregate current emissions from all sectors in the following section.

2.4   ESTIMATING PER AREA CARBON FLUXES FOR GREAT BRITAIN

The carbon fixing and storage characteristics of different forms of land use are summarised in Table 2.2 for 20 land use categories. From the information in Table 2.2, factors for soil and biomass change are calculated. Detailed below is an

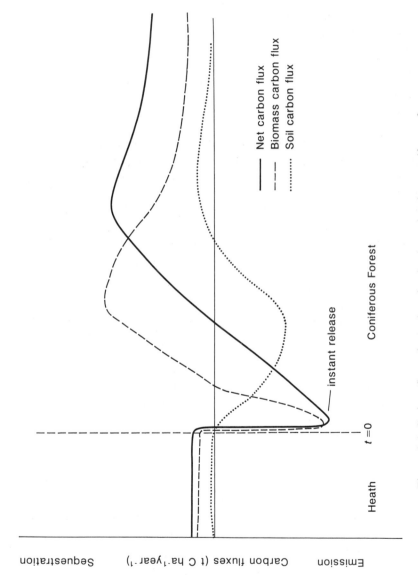

Figure 2.4 Carbon fluxes from changing land use: heath to coniferous forest

Table 2.2 Carbon storage under different land uses

	Additions to soil (t C ha^{-1})	Non-harvested biomass (t C ha^{-1})	Soil type	Equilibrium soil carbon (t C ha^{-1})
Broadleaved woodland	*0–5	0–164	G	170
Coniferous woodland	*0–4	0–95	SHP	450
Mixed woodland	*0–4.5	0–129	GSH	250
Upland heath	0.9	2.4	SZ	200
Upland smooth grass	2.0	2.0	GSH	180
Upland coarse grass	1.3	3.2	HPS	400
Blanket bog	0.7	3.2	P	1200[†]
Bracken	1.5	1.6	SZ	200
Lowland rough grass	2.1	2.4	G	120
Lowland heath	1.0	1.6	Z	80
Crops	2.7	0.0	BG	60
Market garden	1.5	0.0	B	50
Improved grass	3.9	1.6	GB	90
Rough pasture	1.4	2.8	HSP	350
Neglected grassland	2.1	2.4	GS	120
Built up[‡]	0.4	1.2	BGP	10
Urban open space[‡]	1.2	4.0	GBP	70
Transport[‡]	0.4	1.0	–	70
Mineral workings[‡]	0.4	0.8	–	90
Derelict[‡]	0.8	2.0	–	120

*Excluding final harvest waste. [†] No upper limit. [‡] Not in primary land use sector. Soil types (from Avery, 1980): G, stagnogley; S, humic stagno podsol; H, humic gley; P, peat; Z, podsol; B, brown earth.

explanation of the source of the estimates of soil and biomass changes and the assumptions made in calculating the factors, first with reference to non-forestry vegetation; secondly with respect to forestry; and thirdly in relation to the properties of soil and the influences of land use changes on soil carbon.

Non-forest vegetation

Vegetation can be divided into perennial and annual (or biennial) types. The latter leave no standing live biomass as the plant is usually harvested at the end of each year. The majority of arable crops, such as cereals, potatoes and sugar beet are included in this category. The harvested portion will either be exported from the farm, or will be used on the farm for feeding livestock. Some of the exported carbon may return later in the form of industrial products such as sugar beet pulp, which is used for animal feed. It is assumed here that the portion of the crop fed to animals but undigested will be returned to the land and along with the unharvested portion of the crop will be incorporated into the soil. The standing live biomass for non-forestry categories (Table 2.2) is that part left after the harvested production has been removed. The biomass and the soil results have been averaged over a wide range of crops and environments, and are based on experimental results.

In calculations the grassland is assumed to be efficiently grazed, and the manure returned after digestion is complete or, if the grassland is used for meadow, the manure is returned within two years of harvest, but then may be applied to a different field. Lowland herbage is assumed to be 70 percent digestible and upland herbage, 40 percent. While changes in the amount of standing live biomass in lowland grass are assumed to be complete in two years from establishment, a period of five years is assumed for upland vegetation.

Changes in biomass carbon are assumed to occur within a short period for all categories of land use except forestry. It is assumed that changes occur once within the time period. These factors were calculated from data presented in Table 2.2, illustrating the amount of carbon stored under different land uses as standing biomass, excluding the harvested material leaving the sector, as explained above. For the non-forestry category cells where no change in land use has occurred, there is no change in standing biomass from year to year.

The off-diagonal elements for biomass impact factors are the difference between the estimates for non-harvested biomass in Table 2.2 for movements between categories. Although the forestry categories have been assigned a range of values, a single estimate for forestry biomass is taken, as now explained.

Forestry

Carbon storage by trees is more complex than for other crops due to the long life cycle of trees. This section considers the average amount of carbon fixed by forests in Britain. An explanation of the method adopted and assumptions made in the subsequent calculations of factors for biomass change is given. The amount of carbon fixed and stored will depend on a variety of factors including age, species and management. Table 2.3 shows the proportion of conifers and broadleaves of various ages in British woodlands greater than 0.25 ha. Over 77 percent of conifers in Britain have been planted since 1950.

Yields of the various species are estimated using the Forestry Commission Yield Tables (Hamilton and Christie, 1971). Here, measurable volume is described as stemwood greater than 7 cm overbark, and each species is divided into a number of *yield classes*. The yield classes used in the calculations here are those considered most representative, and are therefore taken as equivalent to average conditions. The yield classes used are for conifers, sitka spruce (YC 14), and Scots pine (YC 8); and for broadleaves, oak (YC 6), beech (YC 6), sycamore, ash and birch (SAB) (YC 8) where YC represents growth rate in $m^3 \, ha^{-1} \, yr^{-1}$.

In order to convert the yield in cubic metres per hectare to tonnes of carbon in biomass per hectare the following calculations were made:

(i)   multiply by specific gravity to give dry weight timber,
(ii) multiply by fraction of carbon in dry weight timber,
(iii) multiply by a stemwood to biomass multiplier to include non-harvested
      material (branches and stems less than 7 cm diameter and roots).

These factors vary according to the species and age of trees. The factors used are shown in Table 2.4. The values are weighted according to the proportion of trees of each species in each age group recorded in the woodland census of 1980 (Locke, 1987). Species were then weighted according to the proportion in each age group, as shown in Table 2.3.

When considering the effects of changes in land use, it is assumed that mature conifer woodland is felled at 55 years on average and mature broadleaves at 100 years. These mature forests will have a greater amount of stored carbon per hectare than the averages for the 'national forest', constituted of trees in the age distribution shown in Table 2.3. Even if they are replaced with new plantations consisting of fast-growing species, it will result in a considerable release of carbon dioxide from decaying leaves, branches and roots into the atmosphere, not compensated for by the carbon fixed by new trees. Harmon et al. (1990) have shown how conversion of old-growth forests to young forests results in a net export of carbon dioxide to the atmosphere. Simulations of carbon storage suggest that this holds true even when sequestration of carbon in wooden buildings is considered.

A number of assumptions are made when considering the flows of carbon in British forestry. Mixed woodland is assumed to consist of 50 percent conifers, 50 percent broadleaves. The carbon fixing of thinnings from plantations is ignored as it is assumed that the sequestration rates of thinnings is minimal, and that thinnings are used in the main to replace timber products which are discarded, or which have a very short life. The assumption that thinnings do not add to the stock of stored carbon is the same as that for leaves or needles, in which carbon is

Table 2.3   Percentage of area of trees by planting period in Great Britain

Planting period	Conifers (%)	Broadleaves (%)	All trees (%)
1971–1980	24.9	3.3	18.5
1961–1970	27.7	6.4	21.3
1951–1960	25.3	11.9	20.0
1941–1950	8.8	13.5	10.2
1931–1940	6.7	8.6	7.3
1921–1930	4.6	8.6	6.4
1901–1910	0.9	8.0	3.0
1861–1900	1.5	21.7	7.5
Pre-1861	0.5	11.1	3.7

Source: Adapted from the 1979–1982 woodland census (Locke, 1987).

stored for only a short period. No account is taken of ground cover under the growing trees. Weed control is assumed to be effective at planting (given Forestry Commission recommendations for Glyphosate applications and cultivation; Insley, 1988), and it is assumed that carbon storage in herbaceous vegetation and trees less than 9 years old is very small (see Edwards et al., 1981).

It is assumed that the amount of carbon harvested in usable timber is represented by the Forestry Commission's felling forecasts (Hamilton and Christie, 1971). Table 2.5 shows the carbon harvested as timber, and that left as waste in coniferous and broadleaved woodlands, given the conversion factors presented in Table 2.4. The estimated felled yields given the species mix in Table 2.4 are: 52.6 t C ha^{-1} for conifers, 66.8 t C ha^{-1} for broadleaves, and 59.7 t C ha^{-1} for mixed woodland.

In the present study, a boundary has been set at the farm (or plantation) gate. This means that once timber has left the farm or plantation, the fate of the carbon it contains is not considered. This results in the sector appearing to fix and store quantities of carbon. However, in reality, a substantial proportion of that carbon stored may be released within a very short time. Whilst carbon stored in forest biomass is easily assessed, the fate of that carbon after felling is considerably more complex, but would need to be considered in assessing the effectiveness of forestry for long term carbon storage (see Dewar, 1990, for example).

Table 2.4 Conversion factors—yield to carbon stored

	Sitka spruce	Scots pine	SAB	Oak	Beech
Specific gravity	0.34	0.46	0.51	0.61	0.60
Carbon fraction	0.42	0.42	0.46	0.46	0.46
Stemwood biomass					
Trees < 40 years	1.2	1.2	1.2	1.2	1.2
Trees 41–50 years	1.6	1.6	1.6	1.6	1.6
Trees > 50 years	1.6	1.6	2.0	2.0	2.0
Species weighting	0.686	0.314	0.432	0.397	0.171

Table 2.5 Total biomass and felling yields for broadleaved and coniferous species

	Age (years)	Total biomass		Felling yield		Waste	
		t ha^{-1}	t C ha^{-1}	t ha^{-1}	t C ha^{-1}	t ha^{-1}	t C ha^{-1}
Sitka spruce 14	55	660.8	94.4	365	52.1	295.8	42.3
Scots pine 8	55	353.6	68.3	200	38.6	153.6	29.7
Oak 6	100	522.0	146.5	230	64.5	292.0	82.0
Beech 6	100	594.0	163.9	260	71.8	334.0	92.1
SAB 8	80	534.0	125.3	230	54.0	304.0	71.3

The fate of the waste material—that left behind after felling—will depend on the next use of the land. It is assumed that if the land is to be replanted as woodland, then the waste material will be left on site. This material will decay and the carbon contained within will be released as carbon dioxide via the soil pool over a given period of time. In the case of coniferous woodland, it is assumed that half of the material decays over a period of 12 years, whereas in broadleaved woodlands, decay will proceed more rapidly, with approximately half of the material decaying in 8 years. If the land is converted to agricultural production, it is assumed that all the material will be burnt. Even larger branches and logs from broadleaved trees are likely to be burnt as firewood (it is assumed that this takes place on the farm or plantation).

As the life cycle of the crop is so long, the amounts added to the soil and the standing crop will be more dependent on the age of the plantation than is the case with non-forest vegetation. Table 2.2 therefore shows a range of values for forest but gives means for other vegetation types.

For forestry, which has a higher average age over the period (retained in the same use and hence recorded in the leading diagonal in Table 2.1), the BF^β_i are estimated as:

$$BF^\beta_i = G_i P_i (\lambda_1 L_1 + \lambda_2 L_2 + ... \lambda_z L_z) \text{ for all forestry } i$$

where
BF^β_i = average yearly increase in biomass carbon for land retained in forestry category i;
λ = weight for species mix;
L = average change in standing biomass in age group k,
where $k = 1, 2, ... z$ and

$$\sum_{k=1}^{z} \lambda_k = 1$$

G_i = average specific gravity for forestry category i;
P_i = carbon as a fraction of dry matter for forestry category i.
For land use changes out of forestry it was assumed, in terms of carbon accumulation, that trees were felled at 55 years for conifers and 100 years for broadleaves. The standing biomass of the species weighted forestry categories (BF^β_i) which changed to other uses were estimated as:

- 74.1 t C ha^{-1} for broadleaved woodland
- 33.3 t C ha^{-1} for coniferous woodland
- 40.2 t C ha^{-1} for mixed woodland

The BF^δ_i for these off-diagonal elements are then the difference between this estimate and estimated biomass carbon accumulation in the new use. For land moving into forestry, the biomass factor is a yearly average of simple growth models for the prevailing species in the UK. So the amount of standing biomass of new forests was calculated as:

- 8.04 t C ha^{-1} for broadleaved woodland
- 5.84 t C ha^{-1} for coniferous woodland
- 6.94 t C ha^{-1} for mixed woodland

The difference between the standing biomass carbon content of past and new land uses, as presented in Table 2.2, represents the change in biomass carbon. The values are divided by the transition period (33 years) to give average yearly change factors.

Soils

There is an association between soil type and vegetation making it possible to loosely group together soil type and land use (Table 2.2). However, more than one soil type is associated with a vegetation type, and also several vegetation types are found on a single soil type. When the vegetation type is changed, there may be an associated soil change which involves an increase or decrease in the soil organic matter content (Crompton, 1953).

The soil impact factors (SF_i) are derived from the figures in Table 2.2 (discussed below), which are estimates of the soil organic matter change under each land use considering the wide range of soils and environments on which each land use occurs. In lowland areas additions and losses of carbon are usually in a dynamic balance so that the soil composition is dominated by the mineral fraction. In wet situations in the lowlands, and widely in the uplands, addition of carbon exceeds loss so that there is a long term process of accumulation leading to peat development (Askew et al., 1985). The balance between addition and loss depends partly on the physico-chemical environmental conditions, and partly on the quantity and type of organic material returned to the soil as dead plant tissue (Jenkinson, 1988). Excessive wetness is the major factor limiting peat development, but acidity is commonly an associated factor. Draining and liming can remove both of these constraints. Except where peat develops, it is possible to suggest a carbon content at equilibrium from a wide range of data published by the Soil Survey.

In their natural state peatlands accumulate organic matter and are thus sinks for carbon dioxide (see Chapter 4), but large areas of peat are losing this capacity due to exploitation for fuel and industrial uses, and drainage for agriculture and forestry. Experiments carried out in Finland (Silvola, 1986) show how the carbon balance of peat changes after drainage. Whilst undrained mires were found to accumulate carbon at a rate of about 0.25 t C ha^{-1} year^{-1}, after drainage carbon was released as carbon dioxide at approximately 2.5 t C ha^{-1} year^{-1}. Bouwman (1990a) assumes an initial rate of carbon loss from drained peatland as 10 t C ha^{-1} year^{-1}. Given the longer growing season in Britain and hence longer period over which peat oxidation will take place, the rates of loss would be expected to be higher than in Finland, resulting in a loss rate of perhaps $5 \text{ t C ha}^{-1} \text{ year}^{-1}$ in lowland areas such as the Fenland.

The effect of changing land use between deciduous forest and arable use has been widely studied. At Rothamsted Experimental Station part of the Broadbalk site and the Geescroft site were abandoned in the late 1880s and allowed to revert to woodland. Both sites, which had been in arable cultivation for many years prior to the abandonment, became colonised by a mixture of deciduous species including ash, oak, hawthorn, sycamore, elm, maple and elder. Subsequent changes in soil properties and nutrient contents have been monitored. It was found that over 80 years the soil carbon levels at the Broadbalk site rose from approximately 60 to 110 t C ha^{-1}, and those at Geescroft from 61 to 84 t C ha^{-1} (Jenkinson, 1971). Although there is a substantial difference between the two sites presumably due to differences in the soil properties, the carbon content initially increased rapidly at both sites, but had not yet reached a new equilibrium in the period studied.

When forest is felled there will be an initial drop in organic matter content, and over a long period of time soils will come to equilibrium under new conditions. Edwards and Ross-Todd (1983) examined soil carbon dynamics in a mixed deciduous forest following clear-cutting with and without residue removal. They assumed that the most drastic changes in soil carbon would occur during the first year following harvest due to the physical disturbance of the soil during harvest and increased soil temperatures after canopy removal. The effects of forest clear-cutting in temperate zone forests on atmospheric carbon dioxide concentrations are due to combustion or decay of harvested woody products and decreased production (decreased carbon dioxide fixation through photosynthesis), due to decreased soil fertility resulting from possible erosion and nutrient depletion.

The conversion of grassland to arable will also lead to a decrease in soil organic matter. Bouwman (1990a) reviews a number of studies on changes in soil carbon after changes in land use and concludes that conversion of temperate grassland results in a mean loss of 28.6 percent, ranging from a gain of 2.5 percent to a loss of 47.5 percent. Cooke's (1967) review of British research shows that a third of the carbon in old grassland was lost in 12 years of arable cultivation, while on old arable land 12 years of grass increased the carbon content by 50 percent. As expected, short term grass rotated with arable crops had a much smaller impact on carbon contents.

Soil carbon in mineral soils only reaches equilibrium some 200 years after change in land use (Jenkinson, 1988). Thus, in examining changes in soil organic matter over the 1947 to 1980 period, it is necessary to consider continuing changes which were initiated before the period began, as well as changes which occur within the period. Presumably the latter will also have a larger impact per unit area than will the former. Some idea of the extent or direction of former management changes can be obtained from contemporary comments (Strutt, 1970) while land use changes can be followed from agricultural statistics. In drained lowland peat there is an ongoing loss of organic matter (Hodge et al., 1984) which is counted separately in the carbon balance, while in the uplands peat has spread widely—with fluctuations depending on climate and economic factors. It does appear that an equilibrium is eventually reached in peat deposits, but this is not achieved for several hundred years, during which over 1000 t C ha^{-1} may accumulate (Clymo, 1983).

Due to the complex dynamics of soil carbon properties as discussed above, where soil carbon may be in a continual state of flux towards equilibrium over 200 years and influenced by management practices, fluctuations in soil carbon are simplified into two classes:

(1) Soils under a single land cover moving towards an equilibrium. The variables determining these factors are historically specific such as management practices, straw burning, drainage and liming. These are calculated for each land use category assuming proportions of soil type under each land cover and generalised climatic conditions and therefore the rate of carbon change cannot be applied to a specific location in Britain. The rates are derived from the most widely accepted model of soil organic matter (Jenkinson, 1988) with additional data on soils under forestry. Although no use change has occurred, allowance is made for felling of mature forest and for the slow period of approach to equilibrium resulting from use change before 1947. The resultant factors are the relevant leading diagonal soil impact factors.

(2) Soils under land which has changed use will have an assumed 20 year rapid transition period. It is calculated that, in this period an average of 50 percent of the soil carbon change will occur when simulated for a range of environmental conditions in Britain. The remaining change will occur over a longer period as discussed in the soils under a single land use above. These changes were again based on the Jenkinson (1988) model of soil organic matter.

The rate of change of land use from one category to another over the 33 years is assumed to be at a constant rate. Thus one $\frac{1}{33}$rd of the area in any particular cell in the area matrix changes in the first year, and so on. As the change in land use progresses with time, part of the area will be under the previous land cover, part will be in the 20 year rapid transition phase and part will be moving slowly towards equilibrium under the new land cover. The factors of soil carbon change in the off-diagonal elements are therefore an amalgamation of the two classes of soil carbon change outlined above. With the assumptions of constant change over the 33 years and a 20 year rapid transition period, a mean annual change is calculated based on a weighting of the changes in the separate components in the following proportions:

— slow change in original use	0.5
— rapid transition between original and new cover	0.4224
— slow change in new land use cover	0.0776

From the previous discussion, yearly average factors for each cell in the land use transition matrix are calculated for soil and biomass changes. These factors are then multiplied by the areas to give totals of soil and biomass carbon imported and exported and are discussed below.

2.5   AGGREGATE LAND USE RESULTS

Carbon fluxes are summarised in Tables 2.6, 2.7 and 2.8. The tables show the average per year change in carbon from the soil and from the biomass and the net effect of these, for each of eight broad categories of land use. These categories of land use are merged from the 20 categories in the original database, with the main categories of change, such as forests being counted separately. The biomass results for different land uses, especially woodland, in other studies are often highlighted without reference to the change in carbon content of the soil. The results presented here isolate the pertinent features of the additions and losses from soil and biomass, and the net effects of these.

Table 2.6 shows the average per year change in carbon, given the land use at the end of the period ($\gamma.(BF^\gamma + SF^\gamma)$). This is the average yearly flux in carbon due to the area retained in its original use over the period and the area moving into the particular category. The net changes are disaggregated into changes in soil and biomass carbon in each category.

The results show aggregate annual sequestrations (+ve figures) and losses (or emissions, −ve figures) for eight broad categories of land use. Some 1.32 mt C year^{-1} accumulate in the soil and biomass. This occurs in spite of loss of standing biomass of 0.290 mt C year^{-1}, due to land use change over the period surveyed. This is predominantly from the loss of mature forest and semi-natural upland to

Table 2.6   Total annual changes in soil and biomass carbon in 1980 (γ)

Land use in 1980	Kilotonnes of carbon		
	Soil ($\gamma.SF^\gamma$)	Biomass ($\gamma.BF^\gamma$)	Net
Broadleaved woodland	191	451	642
Coniferous woodland	281	−71	209
Mixed woodland	80	75	155
Crops	−1027	−266	−1293
Improved grass	377	−230	148
Upland semi-natural	1761	−182	1579
Lowland semi-natural	5	−8	−3
Other and urban	−59	−58	−117
Total	1610	−290	1320

Source: Reproduced by permission of Academic Press Ltd from Adger et al. (1992a).
Note: rounding in Tables 2.6, 2.7 and 2.8 may result in discrepancies in totals.

newly planted forest and to arable agriculture. The greatest loss occurs from the soil under arable agriculture, reflecting both soil organic matter loss from ongoing use (0.737 mt C year^{-1}) and from annual increases in the area cropped (predominantly from improved and rough grassland) which contributes 0.290 mt C year^{-1}. The largest accumulation is the annual increase in carbon stored in the soil under upland semi-natural uses, such as bracken, bog, heath and grassland. Surprisingly, the annual average estimate for standing biomass under coniferous woodland is positive (a net emission of 0.071 mt C year^{-1}). This reflects the fact that the forestry estate in the period was immature and, although in aggregate the established forest accumulated 0.0073 mt C year^{-1}, the loss of biomass from the previous land use (such as upland heath and broadleaved woodland) constituted an average loss of 0.144 mt C year^{-1}.

The carbon profile of land remaining within the same category and moving between categories will be different, as explained previously, so the results in Table 2.6 are now disaggregated into these two elements. Table 2.7 shows the soil and biomass carbon characteristics of those areas which have remained in the same land use over the period. The net effect on soil carbon of the area in retained land use is an average fixation of 0.942 mt C year^{-1} over the period, and the annual fixation of biomass carbon sums to 0.570 mt C year^{-1}. This table represents 66 percent of the total area of land analysed.

The ongoing fixation and storage of carbon in maturing trees ensures that biomass carbon changes in retained land use are positive. The biomass figures for non-forestry categories take zero values as the standing biomass at the opening and closing of each period is assumed to be the same. The only negative carbon change resulting from retained land use comes about as a result of arable cropping which continually decreases soil organic matter, releasing carbon from the soil. However, the overall results highlight the benefits of retained land use and conservation of existing land use, if the minimisation of carbon emissions is

Table 2.7   Changes in soil and biomass carbon in 1980 on retained land (β)

Land use in 1980	Kilotonnes of carbon		
	Soil ($\beta.SF^{\beta}$)	Biomass ($\beta.BF^{\beta}$)	Net
Broadleaved woodland	188	439	627
Coniferous woodland	182	73	255
Mixed woodland	63	58	122
Crops	−737	0	−737
Improved grass	383	0	383
Upland semi-natural	1570	0	1570
Lowland semi-natural	3	0	3
Other and urban	2	0	2
Total	1654	570	2224

Source: Reproduced by permission of Academic Press Ltd from Adger et al. (1992a).

Table 2.8 Changes in soil and biomass carbon in 1980 of changed land use (δ)

Land use in 1980	Kilotonnes of carbon		
	Soil ($\delta.SF^{\delta}$)	Biomass ($\delta.BF^{\delta}$)	Net
Broadleaved woodland	3	12	16
Coniferous woodland	99	− 144	− 45
Mixed woodland	17	17	34
Crops	− 290	− 266	− 556
Improved grass	− 6	− 230	− 235
Upland semi-natural	191	− 182	9
Lowland semi-natural	2	− 8	− 6
Other and urban	− 61	− 58	− 119
Total	− 45	− 859	− 903

Source: Reproduced by permission of Academic Press Ltd from Adger et al. (1992a).

regarded as desirable.

The main exports of soil and biomass carbon in the period analysed are then due to changing land use, illustrated in Table 2.8. The data show fluxes due to changes into the particular category, and are found by summing the off-diagonal elements in the column of the category. The results show that there is little net yearly change in soil carbon in land moving between categories (34 percent of total area) as soil carbon losses due to areas converted to arable and urban and other uses are offset by fixing of soil carbon, especially under woodland and upland semi-natural vegetation.

Losses of carbon in biomass due to land use change are of larger magnitude than from soil, totalling 860 mt C year^{-1}. The only positive values are for broadleaved and mixed woodland. Changes in use to crops, improved grassland, coniferous woodland and upland semi-natural vegetation all show considerable loss of carbon in standing biomass. Much of this is due to conversion of areas from mature broadleaved woodland and grassland to arable use.

2.6 AGGREGATE UK EMISSIONS FROM ALL SOURCES

The results of this model of land use change can be extended and analysed in the context of policy and action on the greenhouse effect. First, the fluxes from soil and biomass fluxes form only part of the actual fluxes which come from the land use sector, as management of land requires energy inputs; ruminant animals are significant sources of greenhouse gases; and organic matter can be lost or degraded, for example from peatlands, with land remaining in the same category. Secondly, the fluxes from land use from a country such as the UK are generally ignored in policy responses to the greenhouse effect. Although many studies point to the policy implications of deforestation and land use change on the carbon balance in tropical countries (e.g. Schneider, 1993; Brown and Pearce,

Table 2.9   Carbon emissions and fixations by Great Britain land use sector, 1980

Fixed	Kilotonnes Carbon	Emitted	Kilotonnes Carbon
Crops for consumption	5091	Fossil fuel	894
As meat	297	Liming	348
Livestock products	212	Peat wastage	500
Harvested timber	737	Land use change	903
Retained land use	2224		
Total	8561	Total	2645
			Kilotonnes CH_4
		CH_4 — livestock	1194
		CH_4 – soil	201

Source: Adapted from Adger et al. (1992a).
Notes: Livestock emissions include direct emissions as well as from stored slurry (see Chapter 6).

1994b), the carbon fluxes, and indeed sources of other greenhouse gases, associated with land use do not seem to have entered into the calculus in the UK.

A carbon balance for the Great Britain agriculture and forestry sectors is shown in Table 2.9. Land use and management change account for two offsetting factors in the balance, 2.2 mt C year^{-1} fixed in land retained in the same category (β) and 0.9 mt C year^{-1} emitted as a result of land use change (δ). The most important of the other factors are: the carbon leaving the sector fixed in agricultural products and timber; carbon in methane emitted by livestock, their manures and from waterlogged soils; carbon dioxide emission due to liming; and CO_2 emission due to fossil fuel use. Within the general category of semi-natural vegetation, many changes which affect the quality of the habitat occur. This is especially important in the context of the carbon balance for peatlands, which although widespread in the UK (8 percent of the land area in UK and Ireland; Taylor, 1983) are being degraded through mining for horticultural and fuel use. The trend of direct use of peat may be decreasing as a result of an active campaign against using peat products (see Maltby et al., 1992;Rawcliffe, 1992), but the secondary impact of draining land for other uses continues to reduce the carbon storage capacity. The continuous use of previously waterlogged peatland for agricultural purposes, such as in the Fenland of eastern England, draws down the organic matter, which results in the estimate of 0.5 mt C year^{-1} emitted from peatland in Table 2.9.

The anaerobic conditions of wetlands such as peat bring about methane emissions, which are discussed in detail in Chapter 4. For the balance presented in Table 2.9, an estimate of emissions based on Clymo and Reddaway (1971) (between $1 \, g \, CH_4 \, m^{-2} \, year^{-1}$ (in hummocks) and $7 \, g \, CH_4 \, m^{-2} \, year^{-1}$ (in pools) emitted from deep peat) is used. Taking a mean of $4 \, g \, CH_4 \, m^{-2} \, year^{-1}$ for upland peat (12 400 km^2) and $0.5 \, g \, CH_4 \, m^{-2} \, year^{-1}$ for humic gleys (22 600 km^2), and excluding reclaimed lowland peats, soil methane outputs are equivalent to the emission of 0.201 mt CH_4. In these soils, and in solid manure storage systems, much of the methane produced at depth is oxidised in the aerobic surface zone (Schütz et al., 1990) reducing the potential emissions. The estimates for the other factors, such as methane from cattle are discussed in Chapters 4 and 7. Land use change is simply one element in the human management of land. Mitigation strategies which aim to reduce emissions through, for example, increasing the real price of fossil fuels by taxation will directly impact on the agricultural sector, and have ripple effects in the way in which land is used (Rosenberg and Scott, 1994). However, the role of land use in mitigation strategies is little recognised.

Chapter 1 discussed the reasons why this may be so: land use is a product of historical context and evolution, is presumed exogenous to controls of trace gas fluxes due to the diffuse nature of the biological sources, and the fluxes are difficult to quantify so mitigation success is difficult to assess. The emissions from

Table 2.10 Aggregate emissions of the major greenhouse gases for the UK (1988)

Greenhouse gas	Anthropogenic emissions		Land use/ 'natural' emissions	
CO_2	Natural gas	120.7	Soil	−5.908
(million tonnes)	Coal	272.7	Standing biomass	1.064
	Petroleum	267.5		
	Gas flaring	4.8		
	Limestone processing	6.6		
CH_4	Landfill	0.716	Cattle	0.594
(million tonnes)	Coal mining	1.700	Sheep	0.266
	Natural gas leakage	0.783	Pigs	0.008
			Manure	0.150
N_2O	Nylon manufacture	0.090	Fertiliser use/soil fluxes	0.065
(million tonnes)	Road transport	0.003		
	Other fuel combustion	0.002		
CFC 11[*]	Aerosols	14 295	none	
(tonnes)	Blown foam	4378		
	Refrigeration	1325		
	Air conditioning	325		
CFC 12[*]	Aerosols	14 295	none	
(tonnes)	Foam	4510		
	Refrigeration	2735		
	Air conditioning	325		

Source: Based on Brown and Adger (1993).
Note: Negative emissions signify net sequestration.
[*] CFCs are included here for completeness, though their main impact is on depletion of stratospheric ozone, and they were explicitly excluded from the Climate Change Convention as being a substance controlled under the Montreal Protocol. Recent evidence suggests that the stratospheric ozone depletion caused by the presence of CFCs results in indirect atmospheric cooling (see Isaksen et al., 1992).

land use in Great Britain can be seen in a wider context, which is the UK Government's responsibility under the Climate Change Convention to aim to stabilise their emissions at 1990 levels by 2000. In industrialised countries such as the UK, the emissions from land use are considered trivial. Indeed the UK Government in its strategy does account for the land use fluxes of carbon dioxide, but the policy recommendations in the main concentrate on non-biological sources of carbon dioxide, to the exclusion of methane, nitrous oxide and the other gases. However, the results of land use fluxes show that the actual emissions of CO_2, when the biological sources and sinks are taken into account, are 0.72 percent less at 668 mt C (see Adger and Brown, 1993). When all sources and sinks of greenhouse gases are included—both biological and land use related, and from anthropogenic sources—the role of carbon dioxide is put in perspective. Emissions for 1988, based on estimated energy emissions, and other sources, and on the model of land use described here, are presented in Table 2.10. Carbon dioxide accounts for only 67 percent of aggregate greenhouse gas emissions, when the radiative forcing of the other gases is taken into account.

The UK Government estimates annual emissions in 1990 to be in the order of 160 million tonnes of carbon (mt C), or 587 mt CO_2 projected to rise by 14 percent by 2005 (Table 2.11). The UK Government plans to stabilise emissions include consumption tax (value added tax: VAT) on domestic energy use, increased duty on road fuel, and other regulatory and incentive mechanisms (Department of Environment, 1993). The VAT on domestic heating oil is especially controversial for its regressive impacts on poorer households (see Brown and Maddison, 1993; Giles and Ridge, 1993). The role of land use fluxes is not discussed in the UK government assessment, with agriculture simply being counted with industry as a fossil fuel user. The role of new forest planting in terms of the global carbon cycle is given a brief mention in a recent UK House of Commons Environment Committee inquiry (House of Commons, 1993, para. 91–94), by recommending that planting for new Community Forests in Britain should take the carbon cycle implications into account.

Table 2.11   UK carbon emission projections by final energy consumer (mt C)

Source	1990 mt C	1995 mt C	2000 mt C	2005 mt C
Households	41	39	41	42
Industry/agriculture	56	56	58	61
Commercial/public	24	23	26	30
Transport	38	41	45	49
Total	160	159	170	183

Source: Department of Energy (1991).

The targets for compliance with the Convention, the projected 'business as usual' scenario of growth, and the actual aggregate emissions from the estimates in Table 2.11, along with projected changes which assume the ratio of emissions remain unchanged are illustrated in Figure 2.5. The figure illustrates that the target emission level is actually more generous than it should be, in that CO_2 emissions in 1990 are less than 160 mt C by 1.32 mt C (i.e. 158.7 mt C). The situation is the reverse for non-industrialised countries or countries with large undisturbed natural areas, where the sinks associated with land use are large. The CO_2 emission estimate for Sweden, for example, is reduced by 34 percent when land use fluxes are taken into account. This is due to the large forest area actively sequestering carbon (Rodhe et al., 1991; Adger and Brown, 1993). However, the whole greenhouse gas profile means that the target seriously underestimates the contribution of the UK to the overall greenhouse effect, as shown in Figure 2.5.

The role of land use fluxes to total greenhouse gas contributions raises equity issues on whether systems which allocate responsibility—whether they be based on current or historic emissions—should count all these emissions. The wider the inventories of gases are cast, the more likely that responsibility for reducing global warming will be allocated to developing countries (Brown and Adger, 1993).

2.7 SUMMARY AND CONCLUSIONS

Examination of the carbon dioxide and other greenhouse gas fluxes associated with land use, primarily depends on the availability of land use data. The global coverage of data on land use change is variable. Land use data concentrate usually on productive uses. Data from agricultural production censuses show production and yields of the main crops, but this alone does not suffice, as agriculture may be the predominant but is not the exclusive land use in most countries. Data on forest areas and deforestation rates are notoriously unreliable, as discussed in Chapter 3. The forest area and deforestation debate also highlights the definitional problems in land use data. The analysis of land use change presented here for Great Britain makes use of a dataset which allows the previous as well as the present land use to be accounted for in the estimation of greenhouse gas fluxes. From the analysis it is possible to distinguish generalised land use types, characterised by their respective carbon sequestration and emission properties, and to postulate the effects of changes between these different types of land use in terms of the amounts of carbon stored and emitted.

Grassland is shown to store relatively large amounts of carbon both in standing biomass and soil. However, when account is taken of animals grazing the crop, then methane production ensures that this class of land use contributes to overall carbon emissions. Arable land uses tend to have little standing biomass, reflecting the annual harvest regimes of most crops, and small soil carbon content. Soil organic matter levels are continuing to fall under continuously

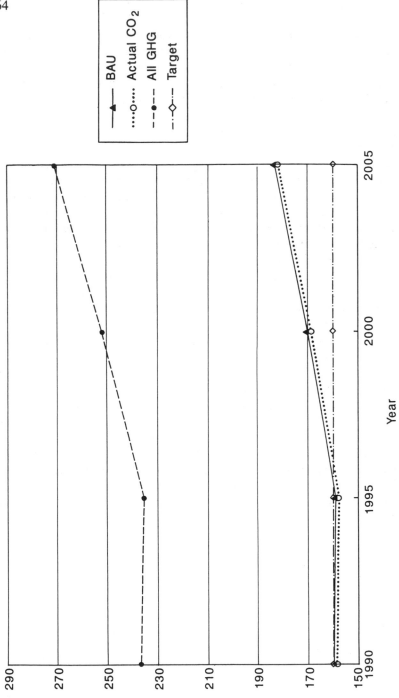

Figure 2.5 UK projected, actual (including land use fluxes) CO_2 and estimated total greenhouse gas emissions, 1990–2005

cropped arable soils towards a new equilibrium. Continued cropping and conversion of land to arable use have been shown to produce the most significant losses of carbon both from the soil and as biomass.

Semi-natural vegetation such as rough pasture, which is not intensively grazed, may store large amounts of carbon, although fixing it at differing rates, depending on age or stage in life cycle, and end use. Blanket bog has been identified as the only land use where carbon accumulation in the soil is expected to proceed for a sustained period. However, large areas are being destroyed, and hence the carbon stored in the soil is being released into the atmosphere as carbon dioxide, as a result of drainage and liming for land use conversion to agriculture and forestry as well as extraction for fuel and for use as horticultural growing media. The only potential for such areas to contribute to the greenhouse effect is through methane emissions, and this is estimated to be small, relative to the fixation in a developing bog.

Although growing trees sequester carbon dioxide from the atmosphere, new afforestation necessitates a change of land use that has effects on soil and biomass carbon which may negate the rationale for afforestation on grounds of carbon sequestration. Grayson (1989) suggests that the afforestation of fast growing plantations in the tropics could offset the current carbon dioxide production of the world's industries. However, as Bouwman (1990a) observes, compared to the atmospheric increase in carbon dioxide from fossil fuel combustion, the sequestration capacity for carbon dioxide by massive plantations of new forests is only modest. In Britain, 0.6 million ha of land is currently being diverted from agricultural land but, as shown in Chapter 7, if this were all afforested with coniferous stands, the UK's net CO_2 emissions would fall by less than one percent.

Data sources on land use change are likely to become better over time as satellite technology improves. The accurate delineation of fluxes associated with land use is one of the factors which has militated against their inclusion in the mitigation strategies. It is by no means the most important reason, however, as feasibility, cost and equity considerations are the key factors in determining whether mitigation takes place.

CHAPTER 3

The Role of Tropical Forests in the Carbon Cycle

3.1 INTRODUCTION

The land use changes of greatest significance to the global climate system are those associated with forests, and in this regard, primarily with loss of forests as a result of conversion to agricultural and other uses. Deforestation is not only one of the major causes of the increasing atmospheric concentrations of carbon and other greenhouse gases, but is among the most significant ecological problems facing the planet. It is the corollary of the other seemingly unstoppable land use change trend, that of the move towards permanent agriculture. In the last half century the predominant site for deforestation has been in tropical regions, whereas previously deforestation of temperate forests was most significant. The present low deforestation rates and net reforestation rates in the temperate zones are explained almost entirely by the low forest cover and the fact that agricultural intensification has taken place in previous centuries (e.g. Mather, 1990).

Fluxes from temperate deforestation exceeded those from tropical regions up to the 1930s, as shown in Figure 3.1 based on the historical regional estimates of Houghton and Skole (1990). Although up to one-third of the increase in atmospheric CO_2 has come about as a result of deforestation, the regions contributing to the biotic flux have changed over time. The expansion of land used for agricultural purposes in the Eurasian and North American land masses was a major cause of deforestation in the 18th and 19th centuries, as well as the demand for wood for development (building and fuel). Although land use change into agriculture continues to the present day (as evidenced for the UK in Chapter 2), the high levels of deforestation have largely disappeared in temperate zones, primarily because only fragments of natural habitat remain, and in some cases are designated as protected areas due to their rarity. However, in the last half century deforestation in the tropical regions has accelerated (Houghton and Skole, 1990).

The anthropogenic sources of carbon dioxide and other greenhouse gases have also increased in recent decades, but according to Woodwell (1992), deforestation

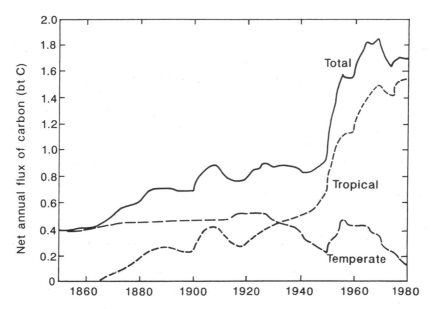

Figure 3.1   Net annual flux of carbon from deforestation in tropical and temperate zones, 1850 to 1980. Based on Houghton and Skole (1990), following Woodwell (1992)

was the dominant source of increasing carbon dioxide in the atmosphere until as recently as the mid-1960s. Between 1850 and 1985 net emissions from deforestation were approximately 100–130 bt C, compared to 190 bt C from fossil fuels in the same 135 year period.

The ecological and environmental consequences of deforestation are not restricted to the emission of greenhouse gases. Local impacts include increased rates of soil erosion; capacity of soils to retain water; other pollutants emitted from biomass burning; loss of biological diversity; and loss of cultural diversity as indigenous people are displaced and indigenous knowledge lost. However, even if the world's remaining forests in both temperate and tropical zones were managed sustainably and deforestation were halted, they are likely to suffer the consequences of climate change (see Chapter 1 and Peters, 1990) and of pollution from other sources (Kauppi et al., 1992). Hence forestry can only form part of the strategy to reduce future climate change impacts, and along with agriculture is one of the principal victims of inaction.

The likely impacts of climate change on the world's forests, as well as possibilities for reforestation of presently degraded lands, need to be considered for an integrated strategy for management of forests to mitigate global warming. We will consider some of the policy issues in later chapters. In the meantime, this chapter examines the carbon fluxes associated with tropical forests and changes in land use from forests. It discusses rates of tropical deforestation, and the methods available to estimate its impact on greenhouse gas emissions.

3.2 GLOBAL DEFORESTATION

Estimation of the role of forests in the global carbon cycle and the greenhouse gas fluxes associated with changes in forest cover involves a certain amount of judgement, and extrapolation of data. The estimates which result do not balance with the data obtained from geochemical analysis of the global carbon cycle; hence scientists postulate that forests may comprise at least part of the so-called 'missing sink' which we referred to in Chapter 1. There is still much debate about the magnitude and precise nature of the missing sink. In order to estimate the greenhouse gas emissions attributable to tropical deforestation the following information is required: information on the type of forests being cleared, and the subsequent land use; estimates of biomass density, and carbon storage in the various types of forest, and in the secondary vegetation; the proportion of biomass which is burned; the proportion of biomass which is kept in long term storage; the rate of CO_2 release in biomass which is neither burned, nor in long term storage.

However, major areas of uncertainty and scientific dispute concern the following: the extent of deforestation in the tropics; estimates of carbon sequestration and storage in vegetation and soils; definitions of vegetation types; subsequent land use and extent; end uses of timber and products. An estimate will enable an assessment of the order of magnitude and the comparative importance of different sources and sinks. It may also be able to highlight the areas of greatest uncertainty and sensitivity which will be helpful in policy formulation. However, sceptics of such modelling highlight the use of the estimates without recourse to uncertainty. As Robinson (1989) comments, balancing geochemical cycles

> is done through guesses and estimates pieced together with surrogation, extrapolation, interpolation, and computation of outer bounds based on physical reasoning. Although they are valuable for demonstrating plausibility and generating interest and debate, such estimates leave science open to unproductive confusion.
>
> (Robinson, 1989, p. 243)

The purpose of the following review is to highlight the uncertainties and the extrapolations required, but still to arrive at caveated central estimates of the impacts of deforestation which inform policy action. Indeed, the existence of uncertainty both leads to economic decisions in favour of preservation (Porter, 1982), and the application of a precautionary approach in policy design to allow for this uncertainty. The first problem is in defining the different types of forest and vegetation cover; the second is in quantifying the extent of deforestation and land use conversion; a third is in identifying subsequent land use.

Classification of tropical forests

The simplest global classification concerning forests is into temperate and tropical. Temperature and precipitation determine the boundaries between major forest zones, but as described by Mather (1990), the world's forests are 'mind boggling' in extent, complexity and diversity. The broad distribution of natural forests is shown in Figure 3.2. The map does not show the actual vegetation cover, rather the type of forest which would exist under natural conditions without human intervention.

This section focuses on forests in the tropical categories in Figure 3.2 (rainforest; moist deciduous and semi-evergreen; and savanna woodland). These tropical forests have rich diversities of species and of forest types, though are frequently distinguished in studies concerning their impacts on global cycles by the extent to which they have been disturbed through human intervention. Myers (1990, p. 373) defines tropical forests as

> evergreen or partly evergreen forests, in areas receiving not less than 100 mm of precipitation in any 6 months for two out of three years, with mean annual temperature of 24 plus degrees Celsius, and essentially frost-free; in these forests some trees may be deciduous; the forests usually occur below 1300 m (though often in Amazonia up to 1800 m, and generally in South East Asia up to only 750 m); and in mature examples of these forests there are several more or less distinctive strata.

Such moist forests exist in more than 70 countries of the tropics, but 34 countries account for 7.8 million km^2, or 97.5 percent of the present biome. Brazil contains approximately 27.5 percent of all tropical forests.

Open and closed forests are defined by the density of trees in the forest. The FAO defines forests as being closed when trees of the different stories and the undergrowth cover a large portion of the ground, and if no grass cover exists. Moist evergreen and deciduous forests account for most of the closed forests. The tree crowns of open forests cover at least 10 percent of the ground surface, which in such forests is typically covered by a continuous carpet of grass. In general, these open forests correspond to dry deciduous forests of the tropics.

A more complex classification system is presented in Table 3.1, showing that even within the predominantly hot tropics, elevation, soils and precipitation lead to a wide variety of species and levels of diversity. The first three are the most important types as far as this discussion is concerned—in terms of their extent, rate of destruction, and carbon-fixing properties.

In addition to classification according to climatic and site-related criteria, forests can also be divided into primary, secondary and logged-over forests depending on their condition. Primary forests are virgin forests whose development has been disturbed only very slightly or not at all by human intervention, so that their physiognomy has been determined exclusively by their natural environment. They are also defined as climax forests, being the final stage of ecological succession. Secondary forest includes all stages of successions which take place on naturally bare land or land that has been cleared. Virgin or natural forest in which trees are felled in a more or less systematic manner and to the extent that the stand structure has been changed, are referred to as logged-over forests. These may be included in the group of secondary forests. Some studies distinguish

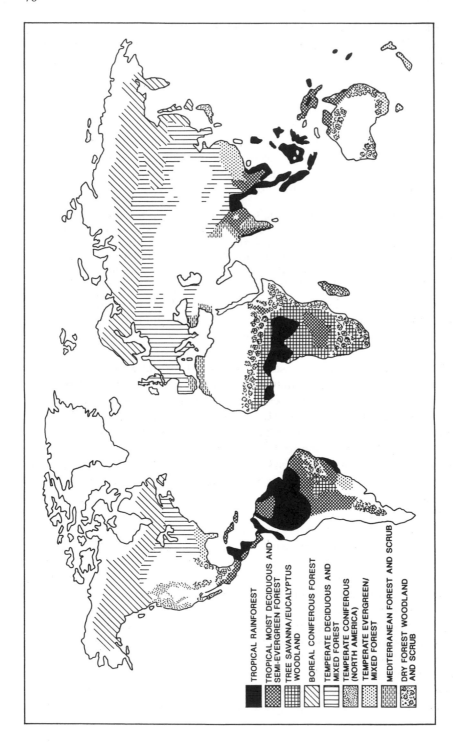

Figure 3.2   Global distribution of natural forest. Reproduced by permission from Mather (1990)

Table 3.1   Classification system for tropical forests

No	Forest type	Description
1	Evergreen moist forest	Equivalent to closed evergreen forest or rainforest, 10° north and south of the equator in Amazon/Orinoco basin, Congo/Guinea basin, parts of India, Thailand/Indochina, and eastern Australia
2	Predominantly deciduous moist forest	Seasonally leafless, monsoonal and transitional. May border the evergreen moist forest with no clear lines of demarcation. This includes all closed types of high forest which shed leaves of at least the upper layer during a clearly defined dry season, with a lower stand density than evergreen moist forest
3	Predominantly deciduous dry forests	Usually found on the edges of type 2
4	Special sites	Including mangrove forests, swamps and flood plains
5	Coniferous tropical forests	
6	Bamboo forests	

Source: German Bundestag (1990).

between productive and unproductive forests. Brown and Lugo (1984) define unproductive forests due to physical reasons (for example, rough terrain or flooding), and legal reasons (for example, national parks or reserves). The area in national parks and reserves accounted globally for only 13 percent of the total unproductive category.

A confusing array of definitions of tropical forests therefore exists, enhancing the difficulty in comparing across studies for forest properties. The production of biomass, and hence of stored carbon is more straightforward than deducing species diversity properties or other ecological criteria across different forest types. However, the uncertainty caused by definitional problems is a significant contributor to difficulties in balancing global carbon budgets.

Estimates of deforestation

Accelerating conversion of tropical forests is occurring for a number of interlocking socio-economic and political reasons (Wood, 1990). These include inequitable land distribution, entrenched rural poverty, and growing populations which push landless and near-landless peasants onto forest lands; government-subsidised expansion into forest regions by plantations growing export crops, timber companies, and cattle ranches; and government-sponsored population relocation to frontier regions. We will discuss these different causes of tropical deforestation with reference to developing policy to reverse them in a later chapter (Chapter Five).

Although forests are a stock of wealth, they are not directly measured over time as part of agricultural or land use censuses. Many deforestation estimates are therefore based on surveys, which in recent years have utilised remote-sensing techniques. Table 3.2 shows the current state of forest inventories in tropical countries: only two out of 40 countries in Africa, and six of 33 in Latin America have conducted more than one inventory, and no assessment was available for three countries in Africa when the survey was carried out as part of the FAO 1990 Forest Assessment.

Historical data are clearly lacking, and even when available, their accuracy may be highly suspect. Grainger (1993) presents a comparison of various estimates made from 1970 to 1990 which highlights serious discrepancies. Table 3.3 illustrates the range in magnitude of these different estimates. First, it is not clear whether the estimates are strictly comparable, as they use different definitions of deforestation—be it total clearance, selective logging, or degradation—and also different types and categories of forest. Second, methodologies and techniques of measurement differ, and remote sensing has done little to provide more accurate data. Remote-sensing techniques offer conflicting estimates, depending on resolution and sampling coverage. For example, two estimates of deforestation in Brazilian Amazonia using remote sensing provide very different figures. One group at the Brazilian National Space Agency (INPE) estimates a rate of deforestation at 1.7 million ha year^{-1} for the period 1978–1988, whereas another group at the same agency estimated 8.1 million ha year^{-1} for 1987. The different figures were produced as a result of using different resolutions, and the second measured smoke from fires as an indicator of deforestation. Accurate estimates wholly based on remote-sensing measurement are not currently available, so on-the-ground monitoring of forest areas or even of stand growth would presently be the only option. Declining rates of clearing have been reported every year since 1987.

One emerging message from the more recent studies is that the rates of deforestation in the tropics have increased during the 1980s, increasing by 50–90 percent over the decade according to the most recent estimates of Myers and of FAO. Much of the difference between the estimates of the increasing rate come from controversy over the single country highlighted above: Brazil. Most of the increase reported by Myers (1989a) from his earlier estimates comes from higher rates in Brazil, which did peak in 1987, with an upper bound estimate of 8 million ha deforested in that year (Setzer and Pereira, 1991). The reasons for ambiguities in the Brazilian estimates (summarised by Houghton, 1993b) mirror the uncertainties for the whole assessment of tropical deforestation: errors of measurement and extrapolation including the interpretation of satellite imagery obscured by smoke from fires; different definitions of the boundaries of the study; and actual yearly differences due to weather conditions or policy changes.

The first global analysis of deforestation was the 1980 FAO/UNEP Tropical Forest Assessment. Completed in 1982, this aimed to provide the first comprehensive and statistically consistent assessment of forest resources, deforestation and afforestation, in tropical countries. Since the completion of the 1980 project, the

Table 3.2 State of forest inventory in the tropics

Region	Number of countries surveyed	Forest area information				Number of countries with forest resources data at national level			
		No assessment	One assessment		More than one assessment	Forest conservation and management	Forest plantations	Volume and biomass	Forest harvesting and utilisation
			Before 1980	1980–1990					
Africa	40	3	17	18	2	7	5	2	4
Asia and Pacific	17	0	1	5	11	9	8	8	7
Latin America and Caribbean	33	0	11	16	6	12	4	4	4
Total	90	3	29	39	19	28	17	14	15

Source: Singh (1993).

Table 3.3   Estimates of rates of deforestation in the humid
tropics (million ha per year)

Source	Date	Period	Total (million ha year^{-1})
Sommer	1976	1970s	11–15[*]
Myers	1980	1970s	7.5–20[†]
Grainger	1983	1976–1980	6.1[‡]
Myers	1989a	late 1980s	14.2
FAO	1990	1981–1990	16.8[§]
FAO	1992	1981–1990	12.2[¶]
FAO	1993	1981–1990	15.4[‖]

Source: Adapted from Grainger (1993).

[*]15 commonly quoted studies.
[†]7.5 a later revision.
[‡]cf. 7.3 for all tropics.
[§]interim estimate.
[¶]revised interim estimate presented to UNCED.
[‖]1990 Tropical Forest Assessment as reported by Singh (1993): figure for all tropics.

need for a continuous process of global forest inventory has been made apparent, and a new assessment of tropical forests was initiated in 1990, intended to be part of an ongoing monitoring and data gathering process as reported in Table 3.3. The 1990 assessment consists of two phases: the first based on collation and analysis of existing information, and the second using remote-sensing (multidate high-resolution satellite) data (Singh, 1993). It also attempts to overcome many of the problems of comparability by issuing guidelines (published by FAO in 1990), and has developed a software programme known as Forest Resources Information System (FORIS) for use on personal computers designed for easy entry, retrieval and storage of data.

Table 3.4 presents data from the 1990 Assessment. This shows that total forest area in the tropics has decreased from 1910.4 million ha in 1980 to 1756.3 million ha in 1990; this represents a decline of 15.4 percent. Annual rates of decrease between 1981 and 1990 are calculated to be 0.7 percent in Africa, 1.1 percent in Asia and 0.7 percent in Latin America; equivalent to a decline of 0.8 percent for all tropical countries. Although comparisons are difficult because of changing definitions, deforestation rates may have been lower, probably running at annual rates of about 0.6 percent, in the late 1970s.

3.3   GREENHOUSE GAS EMISSIONS FROM DEFORESTATION

Despite the difficulties of determining how much of the existing forest is being lost, many studies have attempted to show the impact of this important land use change on the global carbon budget. The most straightforward way to do this is

Table 3.4   Estimates of forest cover area and deforestation by geographical subregions

Geographic regions and subregions	Number of countries surveyed	Total land area (million ha)	Forest area (million ha)		Annually deforested area (million ha)	Rate of change 1981–1990 (% per annum)
			1980	1990		
Africa	**40**	**2236.1**	**568.6**	**527.6**	**4.1**	**−0.7**
West Sahelian Africa	9	528.0	43.7	40.8	0.3	−0.7
East Sahelian Africa	6	489.7	71.4	65.3	0.6	−0.8
West Africa	8	203.8	61.5	55.6	0.6	−0.8
Central Africa	6	398.3	215.5	204.1	1.1	−0.5
Tropical Southern Africa	10	558.1	159.3	145.9	1.3	−0.8
Insular Africa	1	58.2	17.1	15.8	0.1	−0.8
Asia	**17**	**892.1**	**349.6**	**310.6**	**3.9**	**−1.1**
South Asia	6	412.2	69.4	63.9	0.6	−0.8
Continental south-east Asia	5	190.2	88.4	75.2	1.3	−1.5
Insular south-east Asia	5	244.4	154.7	135.4	1.9	−1.2
Pacific Islands	1	45.3	37.1	36.0	0.1	−0.3
Latin America	**33**	**1650.1**	**992.2**	**918.1**	**7.4**	**−0.7**
Central America and Mexico	7	239.6	79.2	68.1	1.1	−1.4
Caribbean	19	69.0	48.3	47.1	0.1	−0.3
Tropical South America	7	1341.6	864.6	802.9	6.2	−0.7
Total Tropics	**90**	**4778.3**	**1910.4**	**1756.3**	**15.4**	**−0.8**

Source: Singh (1993).

to extrapolate using global deforestation estimates, from per unit area estimates of fluxes involved in the changing land use. This is done by estimating the average difference in carbon stocks between forests and the subsequent land use. The following section describes previous global estimates of annual fluxes which have used the FAO 1980 deforestation data. The issues concerning the estimation of the average fluxes from land use are then presented. The non-CO_2 gases associated with biomass burning are subsequently included, though these contain greater uncertainty from being based on site-specific estimates and assumptions as to whether burning is the actual mechanism by which change of use occurs.

Table 3.5 shows a range of estimates of CO_2 emissions from deforestation in 1980. Those of Houghton et al. (1985), Myers (1989a) and Hao et al. (1990) have produced similar ranges, though they disagree on a best guess estimate and on assumptions and methodologies used in deriving the estimates. Hao et al., for example, consider only instantaneous release, rather than net releases associated with regrowth of previously deforested areas and from decay, assuming these to be equal (Houghton, 1991). This assumption is also made for the estimates presented in the following sections. The differences in some of the other estimates are due to the exclusion or inclusion of fluxes from fallow agricultural land, or from calibration of fluxes specific to particular countries.

No global estimates presently exist on annual fluxes based on the 1990 FAO assessment of deforestation, and the global data are not available to estimate the fluxes here. Emissions based on Myers' (1989a) deforestation estimates, such as Houghton (1991), not surprisingly show increased emissions; an increase of 60 percent (877 to 1398 mt C) from 1980 to 1989 in the emissions from closed forests

Table 3.5   Estimates of CO_2 emissions from tropical deforestation in 1980

Source	CO_2 emissions (1980) (billion tonnes C)
Andrasko (1990)	0.4–2.8[*]
World Resources Institute (1990)	2.8
Myers (1989b)	0.9–2.5[†]
Molofsky et al. (1984)	0.6–1.1
Detwiler and Hall (1988)	0.4–1.6
Sedjo (1989)	2.9
Houghton et al. (1985)	0.9–2.5[†]
Hao et al. (1990)	0.9–2.5

Notes: It is not always clear if estimates are net of carbon sequestrated by subsequent land use.
[*]Estimated on additional 0.32 mt C as CH_4 from biomass burning.
[†]Myers' mean estimate is 1.8 bt C whereas that for Houghton et al. is 1.7 bt C.

for a subset of countries which was smaller between the first and second dates (Houghton, 1991).

Carbon dynamics in tropical forests

CO_2 emissions from tropical deforestation are usually calculated by one of two methods: the ecological method and the geochemical balance method. The ecological method calculates net CO_2 emissions from tropical deforestation on the basis of CO_2 sources and sinks in the tropical biosphere. The geochemical method regards the total net emissions of CO_2 from land use change as the upper limit for CO_2 emissions from tropical deforestation, and employs mathematical models to calculate the flux of CO_2 between the biosphere and the atmosphere, and between the oceans and the atmosphere. These models are calibrated by studying the ratio of carbon-13 to carbon-12 and radioactive carbon-14 over a period of time. The results can be used to deduce biospheric carbon fluxes.

This chapter concentrates on the direct measurement of CO_2 source by the ecological method because of its accessibility and simplicity. The ecological method calculates the *net* quantity of CO_2 emissions, i.e. net of subsequent vegetation, on the basis of cleared forest and of secondary vegetation, the density of the biomass of the burned tropical forest, the type of secondary vegetation, the release of CO_2 from the cleared biomass over time and from the soil after deforestation and the fixation of atmospheric CO_2 in secondary vegetation.

It is particularly important to distinguish between the active sequestration of CO_2 by trees as they grow, and the storage of carbon in forest biomass and soils. Overall, forest ecosystems store 20–100 times more carbon per unit area than croplands and play a critical role in reducing ambient CO_2 levels, by sequestrating atmospheric carbon in the growth of woody biomass through the process of photosynthesis. When a forest is cut down, not only does the photosynthesis and therefore active fixing cease, but if the wood and timber is destroyed (most commonly by burning of at least a proportion), then carbon stored by the trees in the past will be released as CO_2. There are still a number of scientific uncertainties, particularly concerning carbon dynamics in 'representative' natural and disturbed tropical forests, and carbon fluxes in tropical soils which may account for one-third of the flux from deforestation.

Biomass

An undisturbed moist tropical forest exhibits net growth for about 100 years after its establishment (Kyrklund, 1990), compared with 30–40 years of rapid growth. After this, as far as carbon is concerned the forest reaches a state of equilibrium where emission at night equals daytime absorption, and dieback equals growth.

Grayson (1989) maintains that existing unmanaged forests contribute no net additional carbon storage since standing biomass remains the same, and carbon fixed by growth is balanced by carbon released to the atmosphere through death and subsequent rotting. If such a forest is left undisturbed for longer periods it will probably become a net emitter of CO_2.

The CO_2 absorption rate is directly proportional to the growth rate. In commercial, even-aged stands of forest it is simpler to estimate incremental growth and in Britain, for example, the Forestry Commission produces yield tables, showing growth rates for most common commercial species (coniferous and deciduous) under a range of different conditions, modelled with an S-shaped logistic curve (see Dewar, 1990). However, estimating the *average growth rate per hectare* for natural tropical forest where a wide variety of species may be present (for example, as reported by Pearce (1989) 1 ha of the Yanomano Forest in Peru was found to contain 283 species of tree; there were only twice as many individuals as there were species) is much more complex. Table 3.6 shows estimates of biomass production in tropical forests compiled from various sources, and from different sample sites within the categories. Again there are problems with definition, and in finding a mean rate which can be applied to a range of forests and conditions.

As regards managed plantation forest, Myers (1990) assumes a working mean figure of 20 t biomass ha^{-1} $year^{-1}$ for growth in tropical areas. As Myers notes, there is much variability in figures adduced for growth rates and yields, according to climatic conditions, soil types, and a number of other factors. Myers then assumes that one-half of plant growth is made up of carbon, and that therefore such a plantation can assimilate 10 t C ha^{-1} $year^{-1}$. However, it is not made clear

Table 3.6   Biomass production in tropical forests

Source	Site	Dry Matter (t ha^{-1} $year^{-1}$)	Carbon (t C ha^{-1} $year^{-1}$)
Bolin et al. (1986)	Tropical rainforest	9.88	4.94
	Tropical rainforest	10.19	5.10
	Tropical rainforest	7.75	3.88
	Seasonal tropical forest	7.20	3.60
	Seasonal tropical forest	7.10	3.55
	Seasonal tropical forest	5.50	2.75
Cannell (1982)	Rainforest, Manaus	15.00	7.50
	Rainforest, Ivory Coast	24.60	12.30
	Rainforest, Ivory Coast	17.17	8.58
	Rainforest, Ivory Coast	12.73	6.36
	Rainforest, Ivory Coast	14.97	7.48
Jordan (1989)	Amazon rainforest, mean	12.66	6.33
	Slash and burn after 3 years	5.26	2.62

Note: Assumes carbon is 50% of dry matter (DM)

what the nature of the plantation is, and one assumes that such a figure would not apply to an uneven, natural stand. Myers notes that eucalyptus plantations in southern Brazil have been found to have an average growth rate of over 30 t ha^{-1} year^{-1}, with occasional top yields of 70 t ha^{-1} yr^{-1} though this is outside the reported range of tropical plantation forestry estimated by Schroeder (1992). Sedjo (1989) applies a universal mean rate of 6.24 t C ha^{-1} year^{-1}, which appears to be closer to the expected rate according to data presented in Table 3.6.

The estimation of standing biomass can be made by two methods: destructive sampling, and from timber volume estimates. Table 3.7 reports various estimates from the Amazon region of standing biomass using different methods. As can be seen, the destructive sampling method, where the biomass from small areas of forest are weighed, tends to give higher estimates which are site specific. The site specificity is illustrated by the range of estimates by INPE on lowland and upland sites within Amazonia. Additionally the methods used in these studies vary tremendously (Fearnside et al., 1993) with the number of sample plots, and particularly with measuring wood density and the critical water content of the wood. For example, the discrepancies between the results of Brown and Lugo (1984, 1992), and Fearnside's estimates (1992a, b) can be explained by the

Table 3.7   Estimates of carbon storage in tropical forests (t C ha^{-1})

Source	Total t C ha^{-1}	Above-ground t C ha^{-1}	Below-ground t C ha^{-1}	Location
Destructive sampling				
Fearnside et al. (1993)	311.0	265.0	46.0	Manaus, Amazonia
INPE (various studies)[*]	168–782	151–726	25.4–116	Low estimates from upland open forest, high estimates from high dense forest, Amazonia
Volume estimates				
Brown and Lugo (1984)	166.0	155.1	21.4	Undisturbed broadleaved forest, tropical America
Brown and Lugo (1992)	205.9	192.6	30.6	Dense forest, Brazilian Amazon
Fearnside (1992b)	320	190.5	48.2	Dense forest, Brazilian Amazon
Fearnside (1992b)	272	162.0	41.0	All forest, Brazilian Amazon

Source: Sources given and adapted from Fearnside et al. (1993).
[*] Reported in Fearnside et al. (1993).

inclusion of minor biomass elements such as roots, and the use of wood volume data from various forest inventory studies.

The Brown and Lugo (1984) estimates used data from FAO detailing stand volumes of forests surveyed in 76 countries, covering 97 percent of the area that lies in the tropical belt. These are categorised into two broad classes: closed forests, where the forest stories cover a high proportion of the ground and lack a continuous dense ground cover; and open forests in which the mixed broad-leaf–grassland tree formation has a continuous dense grass layer and the tree canopy covers more than 10 percent of the ground.

To convert volume data from the usual unit for reporting annual growth (m^3 ha^{-1} $year^{-1}$) to carbon storage units (t C ha^{-1} $year^{-1}$), the following calculation is made:

$$CS = MAI * SG * EF * CC$$

where

CS = carbon sequestration (t C ha^{-1} $year^{-1}$),
MAI = mean annual increment (m^3 ha^{-1} $year^{-1}$),
SG = specific gravity,
EF = expansion factor relating total biomass to volume of usable wood,
CC = carbon content as a proportion of dry matter.

Part of the discrepancy between studies is explained by different assumptions between these seemingly easily determined factors. The ratio of total biomass to usable stem biomass is assumed to be 1.6 for closed forests and 3 for open forests. The density of wood varies over a limited range, and when averaged over a mixed forest the range becomes even smaller. Mid range is assumed by Marland (1988) to be 0.52 t m^{-3}. Houghton et al. (1985) report that the fractional carbon content of wood varies between 0.47 and 0.52 of dry matter (DM) or a wider range in other studies. Most authors, however, assume that 50 percent of DM is carbon. Sedjo (1989) uses the following conversion factors for converting volume of stemwood to tC: 1 m^{-3} of stemwood is equivalent to 1.6 m^{-3} of biomass, 1 m^{-3} of forest biomass (stem, roots, branches, etc.) absorbs 0.26 t C equivalent. Sedjo applies an average figure of forest growth of 15 m^3 ha^{-1} $year^{-1}$ of stemwood, which therefore means that 1 ha of *new* forest will sequester 6.24 t C ha^{-1} $year^{-1}$ (15 × 1.6 × 0.26).

Soils

Soil sources of carbon dioxide include respiration of living biomass and breakdown of dead organic matter. After clear-cutting of forest the first of these will be largely eliminated, but the second will be stimulated by the addition of fresh decomposable organic matter to the soil. Over half of the soil organic matter, and thus the carbon, is contained within the top 40 cm of soil profile. The German Bundestag (1990) cites the average carbon content for closed forests to be 133 t C ha^{-1} in the top 100 cm, and 72 t C ha^{-1} in the top 40 cm. For open forests, the equivalent estimates are 80 t C ha^{-1} in the top 100 cm, and 49 t C ha^{-1} in the top 40 cm. The most useful soil carbon data are those compiled by the German Bundestag, giving the proportions of forested area in each category. These are shown in Table 3.8.

Carbon changes with land use conversion

Tropical forests are exploited by people for a variety of purposes, including timber extraction, shifting cultivation, permanent agriculture and pasture. The causes have variously been identified as slash-and-burn agriculture; commercial timber extraction; cattle raising; fuelwood gathering; commercial agriculture; and additional causes such as large dams, and mining. The causes of tropical deforestation are discussed in Chapter 5. These various land uses differ in their effect on vegetation and soil, and therefore in the amount of CO_2 released when a unit area of forest is converted.

Much of the clearing in tropical rainforests is for shifting cultivation. Johnson (1991) estimates that 64 percent of tropical deforestation is as a result of agriculturalists; 18 percent by commercial logging; 10 percent by fuelwood gatherers; 8 percent by ranchers. However, there are different types of shifting cultivation, which may have different effects on carbon fluxes. Davidson (1985) distinguishes two types of shifting agriculture in tropical moist forests. First, the traditional, low intensity form of shifting cultivation which has been practised for many generations, initially involving the clearing of primary forest but afterwards based on a secondary forest fallow system. Secondly, a more destructive form, in which primary forest is cleared and the land cultivated continuously until it is degraded and then abandoned.

Shifting cultivation is an appropriate form of land use under the prevailing climatic and soil conditions in the tropics providing certain conditions are met:

Table 3.8   Estimates of biomass and soil carbon

Forest category	Area (%)	Biomass (t C ha^{-1})	Above-ground biomass (% of total)	Soil (t C ha^{-1})
Lowland rainforest	11	172	83.3	118
Lowland moist forest	19	185	85.5	88
Dry forest	10	146	—	100
Montane rainforest	14	161	81.3	179
Montane moist forest	14	146	88.2	101
Montane dry forest	32	40	70.0	42

Source: Adapted from German Bundestag (1990).
Note: Dry forests correspond to open forest; all others to closed forest.

(i)   farming must be extensive, with long fallow periods of 12–20 years for most
       fertile soils but 30–100 years for nutrient-poor soils, for example those in
       much of the Amazon basin.
(ii)   plot size must not exceed 1–2 ha, so that it can be readily recolonised by
       surrounding forest vegetation;
(iii) crop mixtures and mixes should ensure maximum ground cover to avoid soil
       erosion.

It seems likely that in many parts of the world these conditions are not being met
and shifting cultivation is now posing a threat to forest resources for a number of
reasons: population growth is exceeding the capacity of existing cropland;
farmers are forced to settle and cultivate unsuitable and poor land; and land
scarcity is further aggravated by ownership and distribution.

Shifting cultivation is typically a cycle, which begins with the burning of a plot
of forest. Some of the large trees are left standing. Food crops are planted in the
ashes and harvested for periods of 1–10 years. The yields generally decline over
time as the forest grows back and soil fertility is reduced. Some of the surface
organic matter is oxidised during the burn and in the earliest years of cropping.
The soil organic matter develops again as the forest regrows. The period of fallow,
after which the forest may be burned and cleared again, may last for 3–80 years
depending on the cultural and environmental conditions.

Based on the FAO data, clearing of forests for the different land uses can be
broken down as follows: shifting cultivation accounts for 40 percent of cleared
primary forest; permanent cropping and cattle ranching 50 percent; logging
(which involves removal of 28 percent of above ground biomass) 10 percent. The
proportions vary from region to region, with 35 percent of forest destruction in
tropical America being attributable to shifting cultivation, 49 percent in Asia, and
70 percent in Africa. In the Americas, 31 percent of forest is cleared for conversion
to pasture for cattle.

Carbon will be released at different rates according to the method of clearance
and subsequent land use. With burning there will be an immediate release of CO_2
into the atmosphere, and some of the remaining carbon will be locked in ash and
charcoal which is resistant to decay. The slash not converted by fire into CO_2 or
charcoal and ash decays over time, releasing most of its carbon to the atmosphere
within 10–20 years. Studies of tropical forests indicate that significant amounts of
cleared vegetation become lumber, slash, charcoal and ash. The proportion
differs for closed and open forests; the smaller stature and drier climate of open
forests result in the combustion of a higher proportion of the vegetation.
Houghton et al. (1987) maintain that over the long term, a constant rate of
deforestation for shifting cultivation will not contribute a net flux of carbon to the
atmosphere. Carbon released to the atmosphere during burning balances the
carbon accumulating in regrowth. However, this is probably no longer the case,
as in recent years the areas cleared annually for shifting cultivation have

increased, the rotation length has been reduced, and the area of fallow forests may also have decreased as fallow lands are cleared for permanent use. All these trends have increased the net release of carbon from tropical ecosystems. Increased burning of root residues and of both burning and decomposition of woody biomass within slash-and-burn agriculture has been hypothesised by Uhl (1987). After five years of succession, 86 percent of the plant mass from the pre-existing forest had disappeared in Uhl's study in Amazonia. Biomass accumulation during this time added only 38 t ha^{-1}. Total carbon stocks at five years were well below half that of the pre-burn forest stocks. Based on measurement of tree growth and litter production, total above-ground production averaged 12.58 t ha^{-1} year^{-1} over the five-year study period, a value almost identical to that measured for mature forest and similar to rates in plantation forestry.

Soil carbon declines when soil is cultivated as a result of erosion, mechanical removal of topsoil, and increased oxidation. Oxidisation is probably responsible for the greatest loss and is the only process which directly affects the CO_2 content of the atmosphere. Again the scale and timing depends on the use after clearing. Detwiler and Hall (1988) estimate that conversion of forest soils to permanent agriculture will decrease the carbon content by 40 percent; conversion to pasture decreases the carbon content by 20 percent; shifting cultivation causes a decrease of 18–27 percent; and selective logging seems to have little effect on soil carbon. This assumes that losses caused by permanent agriculture occur over five years; those caused by shifting cultivation occur over two years; and losses due to conversion to pasture occur within a short time. Approximately 35 years of fallow are required to return to a level found under undisturbed forests.

Carbon in subsequent land use

If tropical forest is converted to pasture or permanent agriculture, then the amount of carbon stored in secondary vegetation is equivalent to the carbon content of the biomass of crops planted, or the grass grown on the pasture (this is explained more fully in Chapter 2). If a secondary forest is allowed to grow, then carbon will accumulate, and maximum biomass density is attained after a relatively short time (as little as 45 years). Table 3.9 summarises the carbon content of soils and biomass in the relevant land uses.

Given the ranges of values and the uncertainties in the role of above- and below-ground biomass, in the estimation of total biomass using different methods, and in the influence of management practice and subsequent use on fluxes, central estimates of carbon fluxes associated with tropical forest loss can now be made. Table 3.10 illustrates the net carbon storage effects of land use conversion from tropical forests; closed primary, closed secondary or open forests to shifting cultivation, permanent agriculture, or pasture. The greatest loss of carbon (the negative figures) involves change of land use from primary closed forest to permanent agriculture. These figures represent the *once and for all*

Table 3.9 Carbon storage in above-ground biomass and soils
under tropical forest and subsequent land uses (t C ha^{-1})

Land use category	Biomass (t C ha^{-1})	Soils (t C ha^{-1})
Closed primary forest	167*	116
Closed secondary forest	85–135	67–102
Open forest	68	47
Forest fallow (closed)	28–43	93
Forest fallow (open)	12–18	38
Shifting cultivation (year 1)[†]	10–16	31–76
Shifting cultivation (year 2)[†]	16–35	31–76
Permanent agriculture[†]	5–10	51–60
Pasture[†]	5	41–75

Sources: Adapted from German Bundestag (1990), Houghton et al. (1987).
* Conservative biomass estimate (see text).
[†] Assumes carbon will reach minimum after 5 years in cropland, after 2
years in pasture.

Table 3.10 Changes in carbon with land use conversion (t C ha^{-1})

	Original C	Shifting agriculture	Permanent agriculture	Pasture
Original C		79	63	63
Closed primary	283	−204	−220	−220
Closed secondary	194	−106	−152	−122
Open forest	115	−36	−52	−52

Source: Brown (1992).
Notes: Shifting agriculture represents carbon in biomass and soils in second year of shifting
cultivation cycle. Negative figures represent losses in carbon in vegetation and soil.

change that will occur in carbon storage as a result of the various land use
conversions.

Other greenhouse gas fluxes from deforestation

The estimated stock changes in biomass given in Table 3.10 have additional
impacts on the atmospheric concentrations of greenhouse gases. These are due to
the emissions of non-CO_2 gases, primarily CO, CH_4 and N_2O (see Crutzen and
Andreae, 1990; Goldammer, 1990) which are active greenhouse gases. As outlined
in Chapter 1, the accumulation of CH_4 in the atmosphere contributes possibly 15
percent to total radiative forcing, while N_2O contributes approximately 6 percent.
CO has no direct global warming effect, but reacts chemically with hydroxyl
radicals (OH·) in the atmosphere and thereby affects the concentration of CH_4.

Biomass burning has, in addition, further impacts on micro-climate through direct albedo impacts on solar radiation and changing the probability of cloud formation. The equatorial regions of rainforest are extremely important in absorbing solar energy and redistributing this heat through the atmosphere. Changes to this mechanism through both the burning process and a reduction in forest area could have unknown future consequences in a warmer world (Crutzen and Andreae, 1990). Biomass burning also induces acid rain in tropical areas and has other impacts on the quality and quantity of water in the area through changes in the hydrological cycle and increased sedimentation.

Attempts to estimate emissions from biomass burning have proved to be extremely complex since the emissions are dependent on many environmental variables such as the burn rate and the ambient temperature and moisture, as well as the ratio of living to dead plant material (see Robinson, 1989, for a review). Table 3.11 gives estimates of various non-CO_2 greenhouse gas emissions from biomass burning, based on experimental controlled-burn situations (Cofer et al., 1991, 1993) (and global normalised estimates for N_2O; Crutzen and Andreae, 1990). The results are given as emission ratios as a percentage of carbon, for unit weight of carbon emitted. The data in Table 3.11 show the percentage extra non-CO_2 trace gas emission in addition to the CO_2.

So for land use conversion out of forestry (as presented in Table 3.10), carbon emissions in the form of CO_2 underestimate the global warming impact. The increased impact can be estimated with recourse to the global warming potential (GWP) of the different gases, shown at the bottom of Table 3.11. These are based on the IPCC 1992 report (Isaksen et al., 1992), and show that carbon monoxide and other non-methane hydrocarbons (NMHCs) have short residence times and have an indirect but positive global warming potential in terms of their impact on

Table 3.11 Ratios of non-CO_2 greenhouse gas emissions from biomass burning and their global warming potentials

| Ecosystem | Emissions as percentage of CO_2 | | | |
	CO (%)	NMHC (%)	CH_4 (%)	N_2O (%)
Grassland/savannah	5	0.5	0.3	0.0018
Boreal forest	9	1.7	0.7	na
Tropical forest	7.1	1.7	0.7	0.0030
Atmospheric residence time	months	days to months	10.5 years	132 years
Direct global warming potential	+ ve	+ ve	11	270

Sources: Emissions ratios based on Cofer et al. (1993) and Crutzen and Andreae (1990). For unit weight of CO_2 emission, the figures show the additional percentage non-CO_2 gaseous emissions. Atmospheric residence times and GWPs from Isaksen et al. (1992).
Note: na = not available.

CO_2 and on tropospheric ozone. The GWPs of N_2O and CH_4 are 270 and 11, respectively, based on a 100 year time integration (see Chapter 1). Ignoring the indirect but unquantified (to date) impact of CO and NMHCs, a central estimate of the global warming impact of CH_4 and N_2O is that, given an emission of carbon (CO_2 carbon equivalent), the global warming impact is an extra 20.9 percent for CH_4 but a less significant addition for N_2O. The emission of CH_4 in the deforestation process is then the most important non-CO_2 gas, with biomass burning estimated as 40 percent of total anthropogenic emissions (see Chapter 1). Estimates of N_2O from burning have been revised downwards by an order of magnitude from a decade before, and burning is now considered a minor source of total N_2O emissions (Crutzen et al., 1979; Watson et al., 1992).

Applying the additional emissions from biomass burning to the carbon fluxes estimated in Table 3.11 increases the global warming carbon equivalent emissions by over 2 percent. This is likely to be an underestimate of the total greenhouse gas impact of biomass burning because of the non-methane hydrocarbons, the carbon monoxide and other minor gases. The adjusted emissions are given in Table 3.12. Biomass burning as discussed above does contribute significantly to the global methane budget, but only a small proportion of the total originates in forest burning. The other major sources are from the burning of grasslands, the burning of straw from agricultural wastes, and the use of fuelwood (e.g. Hall and Rosillo-Calle, 1990; Sathaye and Reddy, 1993).

Uses of felled biomass

The above estimates show emissions of carbon if the biomass is lost at the point of land use change. This is true if burnt, but not if used for timber. When calculating the *net* effects of deforestation on carbon flux, it is important to also consider the end uses of timber cleared from the forest. This may contain the bulk of biomass carbon, and although removed from the growing site, the carbon may not be released to the atmosphere as CO_2; it may remain stored for some time depending on the use and the life of timber products produced. A model

Table 3.12   Greenhouse gas fluxes (CO_2, CH_4 and N_2O) from land use conversion through burning (t C equivalent ha^{-1})

	Shifting agriculture	Permanent agriculture	Pasture
Closed primary forest	208.7	225.1	225.1
Closed secondary forest	108.5	155.5	124.8
Open forest	36.8	53.2	53.2

Notes: positive values denote emissions of Carbon equivalent (with 100 year horizon GWPs; see Table 3.11). Burning is assumed at time of conversion, with emissions of CH_4 and N_2O based on Table 3.11.

developed by Dewar (1990) describes carbon storage in forests and in products removed from vegetation (timber, grain, etc.). This illustrates the relation between carbon storage, vegetative growth, rotation length, and the *carbon-retention properties* of products. This enables management strategies to be examined with respect to their effect on carbon storage (see also Thompson and Matthews, 1989).

An additional dynamic process is that of the subsequent management of areas which are in shifting cultivation. There will be additional releases of CO_2 and other gases if there are subsequent fires. Closed forests therefore lose approximately one-half of the above-ground biomass after three years, and half of below ground biomass after five years when converted to shifting agriculture. The biomass from tropical forests can also be stored for more than a century, in the form of graphite carbon (charcoal) or in timber used for building purposes. For example, Fearnside (1985) estimates that approximately 4 percent of the carbon in cleared above-ground biomass is stored in the form of charcoal (based on measurements from Manaus). Houghton et al. (1987) estimate 2 percent of above-ground biomass is stored as graphite.

The estimations presented in Tables 3.10 and 3.12 show that the largest loss of carbon from both biomass and soils occurs with a change of land use from tropical forest to permanent agriculture. This change also degrades the environment in many other ways, including loss of soil fertility and erosion. Conversion to pasture involves changes of a similar magnitude, though if the emissions of methane from cattle grazing this pasture were included, this conversion may have more serious consequences in terms of greenhouse gas emissions. The conversion of tropical forests to shifting agriculture produces considerable emissions of CO_2, and soil and biomass take many years to recover their carbon store. These results are necessarily generalised, and it must be stressed that empirical evidence from individual experimental sites may provide different data. In particular, different practices of shifting agriculture will affect the carbon flux in differing ways. Some methods of shifting agriculture may be less damaging to the environment than others.

Less damaging cultivation practices are therefore being sought *within* shifting agriculture to minimise its environmental impacts. For example, Southworth et al. (1991) have developed a model which simulates tenant farmer colonisation and its effects on deforestation and associated carbon losses. The model is used to contrast the typical pattern of colonist land use in Rondônia, Brazil, with a system of sustainable agriculture. Sustainable agriculture is simulated such that farmers clear plots of 10 ha each in the first two years of settlement. Annual crops are planted in the first and second years; annual and perennial crops intercropped in the third year so that carbon recovery is initiated. After 10 years of intercropping, the land is left fallow for eight years. The intercropping acts to increase the rate of succession, and no significant carbon loss or gain is assumed to take place during this period.

One other strategy which could be used to maximise carbon sequestration and reduce net emissions of greenhouse gases (if fuel substitution takes place) is the extension of plantations to provide biomass energy (as advocated by Marland, 1988; Hall et al., 1990). However, there are many technical, social and political problems associated with large-scale plantations. Barnett (1992) notes the technical disadvantages of plantation forests, which tend to use non-native, fast-growing species: they support lower biodiversity than natural forests; they do not protect soils from erosion to the same extent; there is the potential of pest infestations; they may lead to loss of nutrients from soil, especially as cut timber leaves the site, which in turn leads to declining productivity. There is still much uncertainty concerning the long term yields from such plantations. There are also likely to be detrimental social impacts from such plantations. Unless planned on the basis of local needs for local benefits they may: have negative impacts on rural and tribal people living in the region; disrupt traditional agricultural practices and associated cultural values and traditions; cause loss of potentially useful varieties of crops and methods of husbandry; foster or increase unemployment; increase land conflict; increase rural poverty and landlessness; increase migration to urban areas.

Industrial plantations in the tropics have already reached the stage of development where high yields of wood are feasible on a large scale. From these plantations, harvested every eight years, as much as $3500 \ m^{-3} \ ha^{-1}$ of wood could be produced over a period of 100 years. This is equivalent to seven-times the CO_2 that would be sequestered by the same area of natural forest established at the same time. More efficient management would include shortening rotations to 5–15 years for hardwoods, and 12–20 years for conifers in the tropics; assuming a yield for hardwoods of $35 \ m^3 \ ha^{-1} \ year^{-1}$, this would mean sequestering $9 \ t \ C \ ha^{-1} \ year^{-1}$. Land use strategies for enhancement of natural sinks of greenhouse gases are discussed in later chapters, notably Chapters 7 and 8.

3.4 GREENHOUSE GAS FLUXES FROM AFRICAN DEFORESTATION—CASE STUDY

Given estimates of greenhouse gas fluxes presented above, which are not calibrated for particular regions, we will now examine the impacts of land use change in one region—tropical Africa—to show the effects on greenhouse gas emissions. The FAO assessment of deforestation, based on satellite imagery, allows the estimation of annual changes but has also generated change maps and change matrices for each sample location. This allows an estimation of greenhouse gas fluxes from land use, using the methodology outlined in Chapter 2 and applied there to UK land use. The land use change matrix associated with African deforestation is used to estimate new greenhouse gas fluxes, based on the volume method, and the additional non-CO_2 estimates presented above.

African deforestation

The land use change matrix for deforestation-related categories of land use between 1980 and 1990 is shown in Table 3.13. Forest cover in Africa, as shown in Table 3.4, is approximately 528 million ha, with deforestation running at approximately 4.1 million ha per year or 0.7 percent per year. The estimates of Singh (1993) are a preliminary report of the FAO 1990 assessment, so are not reported with confidence intervals and are said to await further data input.

Following the notation of Chapter 2, the land use change is presented in Table 3.13 as a square matrix, with the leading diagonal elements showing the land retained in the same category over the time period (β), and the off-diagonal elements showing the destination, in terms of land use category, of land changing use (δ). The aggregate greenhouse gas fluxes associated with these are then per hectare estimates of carbon flux and associated trace gases from biomass burning where this occurs, multiplied by the area of change:

$$TF = \sum_{i=1}^{n} \beta_i \cdot X_i + \sum_{i=1}^{n} \delta_i \cdot GF^{\delta}_i$$

where:

TF = total greenhouse gas flux (t C equivalent per year),

i = 1, 2 ... n land use categories ($n = 9$),

X_i = carbon accumulation in category i (t C ha^{-1} year^{-1}),

GF^{δ}_i = greenhouse gas flux associated with land use into category i (t C ha^{-1} year^{-1}).

As most large positive numbers are located above the leading diagonal, most observed changes are of loss of forest area, or of forest density. Permanent conversion to non-forest uses accounted for 16 percent of the total deforestation, with conversion to short fallow agriculture making up 34 percent of the change. Fragmented forest, a category of land increasing over the decade, occurs as a result of progressive clearing leading to a mosaic of forest and non-forest areas. Fragmented forest is an intermediate stage on the way to permanent agriculture.

According to Singh (1993), this profile of the type of land use change, a summary of which is presented in Figure 3.3, is an indication of spontaneous land use change, caused by population pressures in Africa. Although this seems intuitively plausible, the increasing agricultural use of former forest does not indicate an exclusively population pressure explanation of deforestation. The causal relationships are further explored in Chapter 5 (see also Bilsborrow and Okoth Ogendo (1992) for analysis of how population pressure *contributes* to environmental change).

For the African example, the greenhouse gas fluxes associated with land use change in the nine reported categories are estimated with the results shown in Table 3.14. Simplifying assumptions for this analysis are that all categories except plantation forests (defined as established artificial forests) are in carbon equilibrium,

Table 3.13 Forest cover change matrix for the tropical African region

Classes at year 1980	Area of classes at year 1990 (thousand ha)									Total at year 1980	
	Closed forest	Open forest	Forest + shifting cultivation	Fragmented forest	Shrub	Short fallow	Other land cover	Water	Plantation	(thousand ha)	(%)
Closed forest	16781	382	83	292	10	524	248	—	—	18319	24
Open forest	24	10049	48	371	13	118	397	0	1	11022	14
Forest + shifting cultivation	8	15	557	2	4	52	29	—	—	666	0
Fragmented forest	24	40	1	8089	8	6	294	—	—	8461	11
Shrubs	1	11	—	1	3878	—	164	0	—	4055	5
Short fallow	8	11	10	2	—	2255	53	0	—	2339	3
Other land cover	17	38	11	63	87	34	26452	51	—	26753	35
Water	1	—	—	1	0	3	82	2960	—	3046	4
Plantation	—	—	—	—	—	0	0	—	5	5	0
Total 1990	16863	10546	709	8820	3999	2992	27718	3012	6	74665	100
Percentage of total land area	23	14	1	12	5	4	37	4	—		

Source: Singh (1993).

Note: Totals may appear not to be exact due to rounding.

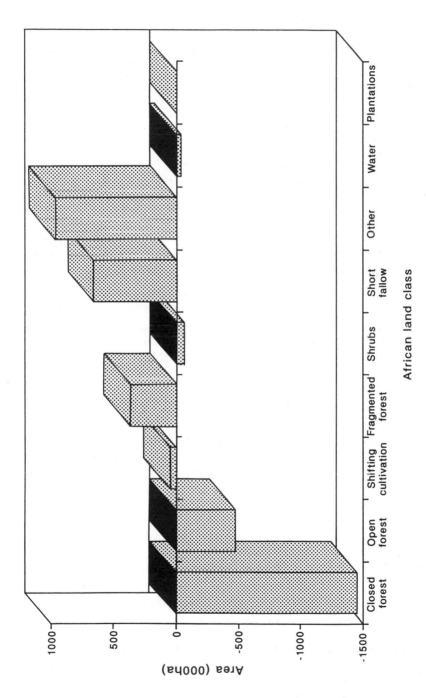

Figure 3.3 Changes in land classes related to deforestation in Africa 1980-1990 (000 ha). Based on Singh (1993)

Table 3.14 Estimated annual greenhouse gas fluxes from deforestation in Africa (mt C equivalent year^{-1})

Land use in 1990	Retained land use, β^* (mt C equivalent)	Land use change into category, $\delta^\dagger$ (mt C equivalent)	Total (mt C equivalent)
Closed forest	0.00	−0.86	−0.86
Open forest	0.00	5.98	5.98
Shifting cultivation	0.00	1.99	1.99
Fragmented forest	0.00	4.09	4.09
Shrubs	0.00	0.17	0.17
Short fallow	0.00	6.36	6.36
Other	0.00	10.55	10.55
Water	0.00	0.00	0.00
Plantations	−0.04	0.00	−0.04
Total	−0.04	28.29	28.25

Notes: positive values denote net emissions, negative values denote net sequestration. Flux estimates are discussed in the text and based on Table 3.12.
*Retained land use (β) denotes land in the same category at the beginning and end of the period (1980 and 1990). A simplifying assumption is made that this land did not move out and then return to the same category during the period (see Chapter 2).
†Changed land use (δ) denotes land changed into the category during the period 1980 to 1990. Units are million tonnes of greenhouse gas emissions (CO_2, CH_4 and N_2O) in carbon equivalent, based on direct global warming potentials of the non-CO_2 gases.

so that β_i for all non-plantation classes in Table 3.13 are zero. The fluxes associated with other land use changes are extrapolated from Table 3.14.

To estimate the fluxes associated with plantation forestry, and hence both the β elements and the net fluxes from the non-diagonal elements into and out of plantations, it is necessary to again simplify carbon sequestration on the basis of estimated volume productivity as outlined earlier in the chapter. The annual sequestration rate for plantation forestry in Africa is based on a mean annual increment (MAI) of 20 m^3 ha^{-1} year^{-1}. MAIs of various tropical plantation species have been assembled by Schroeder (1992) (see also Myers, 1989b; Dixon et al., 1993a) who shows a range from 6 m^3 ha^{-1} year^{-1} for infertile or degraded sites to 25 m^3 ha^{-1} year^{-1} for some pine and acacia species. To convert to carbon sequestration, the MAI (20 m^3 ha^{-1} year^{-1} taken as a central estimate) is converted to carbon sequestration (CS) as discussed above by:

$$CS = MAI * SG * EF * CC$$

With a specific gravity (SG) of 0.5, a carbon content of dry matter (CC) of 0.5, and a stemwood to volume expansion factor (EF) of 1.6, this translates to 8 t C ha^{-1} year^{-1} for African plantations.

Changing land use into or out of plantations requires estimation of average standing crop, estimated again from the annual carbon sequestration (CS), and making the simplifying assumption that all the stored carbon is returned to the atmosphere after harvest. As the greatest proportion of plantation forestry is used for fuelwood, this gives a realistic first approximation. So, following Schroeder (1992), assuming that the representative plantation has an MAI of 20 m^3 ha^{-1} year^{-1} and is hence acacia or pine with an average rotation of 20 years:

$$Average\ CS = \frac{\sum_{i=1}^{rot.} CS_i}{rotation\ length}$$

$$= 84\ t\ C\ ha^{-1}$$

Given these assumptions, the resulting greenhouse gas fluxes from deforestation and related land use are presented in Table 3.14. This gives an overall estimate of 28 mt C equivalent greenhouse gas emissions per year from deforestation in Africa. The only offset is the estimated sequestration in plantation forests and in the changing land use into closed forest from previously cultivated land. The overall assessment is higher than the estimate by Subak et al. (1993), which includes biomass burning in savannas and non-forest land uses not included in the estimate presented in Table 3.15, and is significantly higher than the WRI (1992) estimate of 1.5 mt C emissions from land use for all of Africa (again the estimate here from the 1990 FAO deforestation assessment is for tropical Africa only). The estimate is likely to be higher because of the provisional nature of the deforestation data; the assumption of the instantaneous loss of most of the stored carbon from land use changes out of the closed forest category; and the strict assumption of instant release of CO_2 from use of plantation forests.

The immediate processes which lead to the changing land use are predominantly agricultural in nature, and include accelerating rates of shifting cultivation with shorter fallow periods, as evidenced from Table 3.13, but the underlying causes are in the breakdown of tenurial systems, lack of non-agricultural employment opportunities, as well as the scale of population change (Knerr, 1992). The results presented here show how subsequent land use and management of forests and other resources are important in judging the role of land use in contributing to global warming and for identifying the potential solutions to the land use problem which also impact on the greenhouse effect.

3.5 CONCLUSIONS

Forests sequester and store large amounts of carbon, and the amount of carbon locked up in the biomass and soils under forests exceeds that of most other forms of land use. The loss of forests and the change in land use to other uses, whether agriculture or urbanisation, results in significant emissions of CO_2 and other greenhouse gases. The exact magnitude of the flux will depend on the method of

Table 3.15   Tropical deforestation, biotic sources and all sources of greenhouse gases

		Annual emissions	% of total emissions	Percentage contribution to greenhouse effect in 1980s	
				Total*	From deforestation
CO_2				50	
	Industrial	5.6 bt C			
	Biotic†	2.0–2.8 bt C			
	Tropical Deforestation	2.0–2.8 bt C	26–33		13–16
CH_4				20	
	Industrial	50–100 mt C			
	Biotic†	320–785 mt C			
	Tropical Deforestation‡	155–340 mt C	38–42		8
N_2O[d]				5	
	Industrial	<1mt N			
	Biotic†	3–9 mt N			
	Tropical Deforestation	1–3 mt N	25–30		1–2
CFCs				20	
	Industrial	700 000 t			
	Biotic†	0	0		
TOTAL				95	22–26

*The greenhouse gases considered in this table are only those released as a direct result of human activities. Tropospheric ozone, formed as a result of other emissions, contributes about 5 percent to the total.
†Biotic emissions include emissions from tropical deforestation as well as natural emissions.
‡Relatively little of this CH_4 is emitted from deforestation; most of the emissions result from rice cultivation or cattle ranching, land uses that replace forest. Additional releases occur with repeated burning of pastures and grasslands (see text).
Source: Houghton (1993b).

forest clearance, how biomass is disposed of (most importantly, whether it is burned) and consequent land use. Up until the 1930s, deforestation in the temperate zones was marginally more important, but in the last 60 years the extent and therefore effect of deforestation in the tropics has become the most significant land use contributor of greenhouse gases to the atmosphere. Precise estimates of the exact magnitude of its contribution are, however, difficult to assess: first, because until recently, accurate data on the rate of deforestation in tropical countries were difficult to obtain; and secondly, because we still do not have enough information on carbon fluxes from different types of forest and how these change with conversion to different land uses, such as shifting and permanent agriculture. It is therefore necessary to extrapolate from results obtained from experimental plots, and use our judgement to make generalised estimates of the effects of broad categories of land use change from specific examples. Databases are, however, expanding, and the rate of change in terms of the information available is rapid. We have used the latest available data from the 1990 FAO Tropical Forests Assessment to calculate fluxes of greenhouse gases from land use change in Africa in the decade 1980–1990. This estimates a mean of 28 mt C equivalent emitted per year; this compares with a figure of 176 mt C equivalent of CO_2 emission from industrial sources for the African continent.

Tropical deforestation may have been responsible for between 22 and 26 percent of the greenhouse effect during the 1980s, as shown in Table 3.15. Of global CO_2 emissions, 26–33 percent may even result from tropical deforestation (but given the caveats discussed in this chapter). Tropical deforestation also entails a number of other costs to the environment and to local human populations, and evidence suggests that, despite the fact that the rate of destruction of tropical forests has been decreasing since the late 1980s, it is still at a rate that is environmentally, economically and socially non-optimal. Measures are needed which will lower the rate of deforestation and thus bring a range of benefits, including lowering emissions of greenhouse gases associated with soils and wetlands.

CHAPTER 4

Soils, Bogs and Wetlands: Greenhouse Gas Fluxes

4.1 INTRODUCTION

In this chapter the mechanics of how soils and wetlands interact with the global biogeochemical cycles are described. Wetlands have an inverse relationship with these cycles to non-saturated soils. Saturated natural wetlands have an almost unlimited capacity for accumulation of organic matter, and hence act as a major sink for carbon. This contrasts with dry soils where the organic matter is relatively fixed, and can be drawn down with cultivation. Wetlands also comprise a large source of methane emissions, whereas other soils act as a sink. Future land use change as well as future climate change could radically alter the current set of fluxes. Indeed it is this set of fluxes which constitutes some of the 'unpleasant surprises' and uncertainties about the process of greenhouse gas induced climate change (Lashof, 1989). A major focus of this chapter is the methane emissions associated with the soil part of the biosphere. Wetlands constitute the largest natural emission of methane. The only human influence on this natural system is intervention which causes a change in the scale of the wetland area, rather than in any direct management practice. Artificial wetlands, such as for rice and coarse fibre production by contrast are the single most important *direct* agricultural emission (as opposed to indirect emission through land use change). The policy implications for mitigation of these emissions are highlighted in Chapter 6, the scale of the emissions and the mechanisms by which these occur being the focus in this chapter.

Soils contain organic matter, they cycle carbon through the distribution of above- and below-ground biomass, and they act as a natural sink for methane. Many wetland areas are seasonal so are only under anaerobic conditions for part of the year. The distinction between wetlands and non-saturated soils is fuzzy as soils may be seasonally saturated, and so the categories overlap. For the purposes of this study we define the following broad categories with contrasting dynamics in the global biogeochemical cycles:

— *Wetlands*: soils which are saturated for large parts of the year and hence allow methanogenesis in anaerobic conditions.

— *Mineral soils*: soils which are not saturated for lengthy periods, making up the remainder of the world's soil types. These vary widely in their characteristics, fertility and nutrient content.

An example of the overlap between these categories is that of peat which is characterised by a high organic matter content. Active peatbogs accumulate carbon due to their waterlogged nature and the decaying live material on their surface. Mature bogs do not actively sequester carbon from the atmosphere, having reached an equilibrium state. Peaty soils by contrast are defined by having a high organic content and may have previously been active peat deposits. Presently, however, they are aerobic, fertile soils and in our classification are in the mineral soils category.

Wetlands are especially important as they have larger proportions of organic matter and they tend to be natural or semi-natural, although there are wetlands created specifically for agricultural purposes. They are among the most fertile and productive ecosystems in the world, and are diverse species habitats. Indeed the boundary between land and water has constantly been the scene of major evolutionary processes. The pressure on this set of habitats from human development is great. But their significance is recognised under the Ramsar Convention and they are thus given a protected status internationally. This convention defines wetlands as:

> areas of marsh, fen, peatland or water, whether natural or artificial, permanent or temporary, with water that is static or flowing, fresh, brackish or salt, including areas of marine water the depth of which at low tide does not exceed six metres.
>
> (Ramsar Convention)

Under anaerobic conditions, wetlands produce methane emissions, natural wetlands accounting for one-fifth of the total fluxes to the atmosphere at present, and even larger proportions in past climatic conditions. There is generally no limit to the depth to which organic matter accumulates in wetlands such as peatbogs. The continued expansion of agricultural land around the world leads to a draining of wetlands, representing a loss of habitat for numerous species and also with implications for the global biogeochemical cycles. If a natural wetland is drained, methane emissions are reduced, but stored organic matter is lost through oxidisation to the atmosphere. There is therefore an inherent trade-off in these cycles which is outlined in Section 4.3. Indeed a conference on the role of soils in the greenhouse effect even recommended that wetland drainage should be considered as a priority for scientists and decision-makers (Bouwman, 1990c, p. xvii).

Rice cultivation provides the staple food for over half of the world's population. The cultivation of rice under anaerobic conditions accounts for one-sixth of the anthropogenic methane emissions globally and the largest area of created wetland. Management issues in rice cultivation do affect the efficiency of methane production, so there is scope for future emission reduction. So-called

Green Revolution technologies, which facilitated increases in food production in Asian countries in the past three decades, also have implications for methane emissions.

The geographic distribution of artificial and natural wetlands is shown in Figure 4.1. The largest wetland areas are natural cold wet boreal regions between latitudes 50 °N and 70 °N. However, extensive areas also occur up to 15 ° either side of the equator, associated with the huge swamps and floodplains of South America and tropical Africa.

4.2   CARBON ACCUMULATION IN SOILS

Estimates of the total size of the global biomass carbon range from 600 to 835 bt C (Bouwman, 1990a, p. 62) (compared to 720 bt C in the atmosphere; 38 000 bt C in oceans and 6000 bt C in fossil reserves; see Chapter 1). Soils form a much larger sink than above-ground biomass. It is because this sink is less disrupted by anthropogenic activity, that the major fluxes of carbon are associated with land use change and loss of above-ground biomass. However, as shown in Chapter 2, soil fluxes do occur when land use change occurs.

Carbon organic matter in soils is made up of plant debris or litter in various stages of decomposition, and in the form of humus. Of the total soil carbon, plant litter (the intermediate stage in forming soil) may form 70–80 percent. A large proportion of the soil carbon is therefore found in the first 50 cm below the surface and accumulates over time at a rate dependent on decomposition of the plant litter. The proportion of carbon decreases with increasing depth, but a useful rule of thumb is to assume 50 percent carbon content by weight of soil. The accumulation of soil carbon depends on precipitation, available water, soil clay content and on the vegetation type. For example, under temperate forests soil accumulates to an equilibrium 50–60 years after planting (e.g. Grigal and Ohmann, 1992; see Chapter 2). Although the soil carbon represents a major pool, the greatest dynamic fluxes are associated with those soils which are permanently or temporarily saturated with water.

4.3   WETLANDS

Wetlands cover 6 percent of the world's land surface and are found in all climates from arctic tundra to the tropics. Wetlands provide humans directly and indirectly with ranges of goods and services: staple food plants, fertile grazing land, support for coastal and inland fisheries, flood control, breeding grounds for numerous birds and fuel from peat. 'Natural wetlands' is a collective term for a number of habitats which have come about through different processes, and hence have distinct characteristics with regard to biogeochemical cycling and to change under changing ambient conditions. 'Wetlands' describes ecosystems where formation has been dominated by water, and where processes and

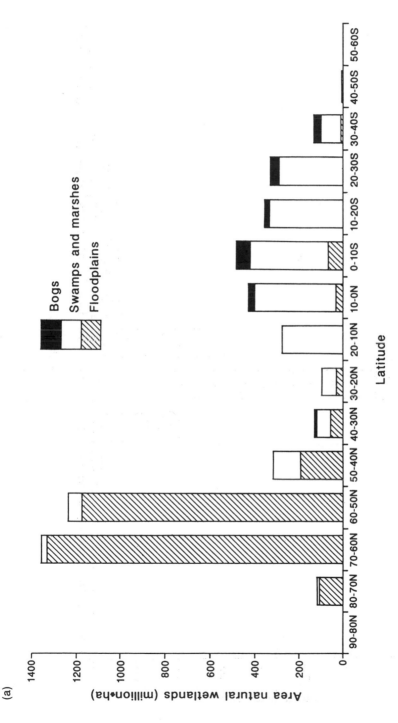

Figure 4.1 Global distribution of (a) natural wetlands and (b) rice paddies along 10° bands. Source: Matthews et al. (1991), Matthews and Fung (1987)

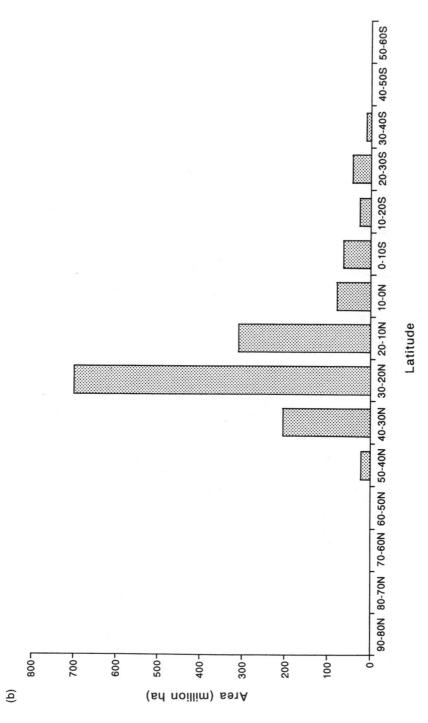

Figure 4.1 *Continued*

characteristics are largely controlled by water.

Wetlands are formed by a number of human and natural processes. The major natural factors are sea level change and heavy precipitation allied to previous geological activity such as depressions left after glaciation. Northern tundras which are wet for the months when they are not frozen occur due to the water being trapped above permafrost. Previous human activity has resulted in wetland formation such as the Norfolk Broads in eastern England which were dug out for their peat deposits. The most significant human-created wetlands at present are rice paddies.

The different types of natural wetland can generally be classified either as marshes, swamps or peatbogs, and their principal characteristics are shown in Table 4.1. Coastal marshes are predominated by either saltwater or freshwater species whereas freshwater marshes are caused by frequent flooding of surface fresh water. It is the flowing of water into the marsh which distinguishes this type

Table 4.1 Characteristics of wetlands

	Principal types	Characteristics	Fluxes
Marshes	Freshwater marshes	Areas kept wet by freshwater sources such as springs and groundwater	Tropical/ temperate
	Saltmarshes	Marshes lying between high and low tide in estuaries, with salt-resistant species; accumulate through sedimentation as opposed to accumulation	Tropical/ temperate
Swamps	Mangrove swamps	Swamps are formed by stationary rather than flowing water and are often flooded forest; 14 million ha globally of mangrove swamps; flooded forest dominated by some of 60 mangrove species	Tropical
	Herbaceous swamps	Reed swamps occurring in both temperate and tropical countries	Tropical/ temperate
Mires	Blanket mires	Rainfed mires generally in climate with greater than 1000 mm year^{-1}. 10 milion ha worldwide	Temperate
	Raised mires	Deep mires often with domes; can be 3 m thick; normally in lowlands	Temperate
	Tundra mires	Usually a thin layer of peat over permanently frozen ground trapping water; occur primarily in Canada and Siberia and Scandinavia with 11–16 million ha worldwide	Temperate

Sources: Maltby (1986), Maltby et al. (1992).

of wetland from bogs, which are essentially rainfed. Swamps are generally areas of still or stagnant water such as on the edges of lakes and on the floodplains of rivers. These may have been formerly forested so may have the appearance of flooded woodland, as in the cypress swamps of the southern United States. Mangrove swamps, dominated by the 60 species of mangrove tree, are the tropical and sub-tropical equivalent of the saltmarshes of temperate climates in that they are intertidal, and according to Maltby (1986), support over 2000 species of fish, invertebrates and epiphytic plants, making them economically important ecosystems.

The organic dynamics of wetlands are complex and especially so for the mire-type habitats which are active in accumulating organic matter. Areas with peaty soils may not be actively accumulating organic matter but still act as a store for previously accumulated organic matter. The major fluxes from mires are of CH_4 and CO_2. From other wetland types, by contrast, as shown in Table 4.3, the major emissions are of CH_4 only, and CO_2 is not actively sequestered. The following section reviews the ranges of estimated emissions from different natural wetland types. Tropical wetlands are the most productive, but the comparatively large area of the temperate and high latitude habitats means that they are also significant in the global fluxes.

4.4   METHANE FROM NATURAL WETLANDS

Methane emissions from wetlands are difficult to estimate, as they are generally extrapolated across a wetland classification from experimental observation of a small area. The fluxes associated with individual sites vary spatially with climate, degree of saturation and seasonally with precipitation. Global estimates therefore have wide variation in their results because of their extrapolation (Bartlett and Harriss, 1993). Tropical wetlands produce CH_4 at a prodigious rate, and although the underlying process of anaerobic methanogenesis is the same worldwide, the wetland types in tropical areas are significantly different to those in temperate areas.

Temperate wetland methane sources

The range of emissions of methane from temperate and boreal wetlands is large, not least because the definitions of wetland types are vague, and because the experimental work estimating emissions uses a variety of methods and timescales. Indicative are emission rates for various wetland types as shown in Table 4.2. These show that fluxes range from 2.0 mg CH_4 m^{-2} day^{-1} for saltmarshes to greater than 100 mg CH_4 m^{-2} day^{-1} for bogs and swamps. Boreal bogs produce even greater levels of emissions. The aggregate emissions from these latitudes are based on the area and average flux assessment of Bartlett and Harriss (1993).

No assessment to date has linked these emissions from natural wetlands to all

Table 4.2 Temperate and boreal methane emissions by wetland type

Type	Temperate (t CH_4 ha^{-1} $year^{-1}$)	Boreal (t CH_4 ha^{-1} $year^{-1}$)	Total emissions mt CH_4 per year (temperate + boreal)
Bogs	23–106		
forested		88–102	23.62
non-forested		177–402	9.80
Forested swamps	113		2.16
Cypress swamps	9.9–39.8		3.16
Saltmarshes	2.0–5.0	3.1	0.31

Sources: Based on Bartlett and Harriss (1993) and Franken et al. (1992).

Table 4.3 Tropical natural wetland emissions

Habitat	Estimated average Amazonian fluxes[*] (t CH_4 ha^{-1} $year^{-1}$)	Estimated tropical average fluxes[†] (t CH_4 ha^{-1} $year^{-1}$)	Area[‡] (million ha)	Aggregate fluxes[§] (mt CH_4 $year^{-1}$)
Open water	0.179	0.540	73.6	30.88
Grass mats	0.737	0.850		
Flooded forest	0.365	0.602	90.5	26.88
Tidal marshes		0.664	15.2	4.98
Forested bogs		0.602	8	2.38
Non-forested bogs		0.695	1.2	0.41
Total				65.51

Sources: Based on Bartlett and Harriss (1993), sources below and own calculations.
[*]Average Amazon fluxes are weighted average of wet and dry season fluxes (Bartlett et al., 1990; Devol et al., 1988) and of the observed annual flux from one available site (Devol et al., 1990).
[†]Tropical average is weighted by results of estimates from other regions such as the Orinoco and Congo basins (Tathy et al., 1992, for example). Annual emissions for bogs are based on averaged daily fluxes of forested and non-forested swamps, with tidal flats emission estimate being an average of those for all swamps (Bartlett and Harriss, 1993).
[‡]Area data based on Matthews and Fung (1987).
[§]Annual aggregate fluxes are based on 180 days wet season emission period.

country areas, although particular country studies have identified emissions, for example Svensson et al. (1991) for Sweden. This study reports present emissions of 2.25 mt CH_4 emitted per year from Swedish bogs, and further suggests that if human-induced fluxes alone were to be counted, then a negative emission of 0.3 mt CH_4 could be taken as the annual emission avoided through wetland draining (Rodhe et al., 1991; Svensson et al., 1991). This confuses the issue of current and historical emission patterns. The data on wetland location summarised in Bartlett and Harriss (1993) would allow a country-by-country estimate of emissions from natural wetlands to be devised, but as the discussions on Swedish wetland emissions indicate, this would not be helpful in designing present and future mitigation strategies.

Tropical wetland methane sources

The major emission sources from tropical wetlands are from mangrove and herbaceous swamps and from open water, which act like permanently inundated swamps with macrophyte species producing CH_4. Although there have been a plethora of studies of tropical wetland emissions in recent years (summarised by Bartlett and Harriss, 1993), these have been concentrated in the Amazon floodplain with a few studies of African herbaceous swamps. Bartlett and Harriss (1993) extrapolate to the whole of the tropics based on their survey of the recently reported studies, using a classification of mangrove and other forested swamps, open water (in effect permanently flooded freshwater swamps) and grass mats (see Table 4.3). Grass mats are formed by the vegetation in seasonal freshwater swamps breaking away from submerged soils and floating freely in large mats on the surface. They occur significantly in the central Amazon floodplain.

Tropical areas then represent over 60 percent of the world fluxes of methane from natural wetlands (shown in Table 4.4). This is a result of both the large areas of swamp and floodplain in the Amazon and African tropical zones, and the high emission rates from these areas. Figure 4.2 shows the relative significance of methane fluxes from natural wetlands by latitude. Future prospects of emissions from these sources are discussed below.

4.5 CARBON IN PEAT AND WETLANDS

Although the natural process of CH_4 emissions from wetlands is reduced through the draining of wetlands, the conclusion of studies such as by Maltby et al. (1992)

Table 4.4   Global natural wetland sources of methane

Climate zone	Wetland type					
	Forested bog (mt CH_4)	Non-forested bog (mt CH_4)	Forested swamp (mt CH_4)	Non-forested swamp (mt CH_4)	Alluvial (mt CH_4)	Total (mt CH_4)
Arctic +non-wetland	8.9	4.9	0.1	0.2	—	14.1
Arctic source (dry tundra)	—	—	—	—	—	4
Boreal	12.6	4.9	0.5	1.4	—	19.5
Temperate	2.1	—	1.6	1.5	0.3	5.4
Tropical	2.4	0.5	26.9	30.9	5.0	65.6
Total	26.0	10.3	29.1	34.0	5.3	108.6

Source: Bartlett and Harriss (1993).

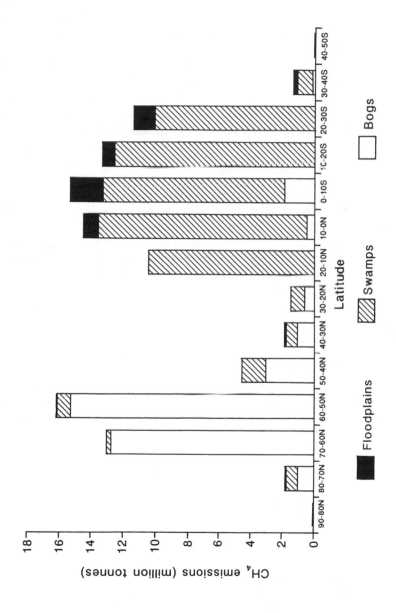

Figure 4.2 Global methane emissions from natural wetlands by 10° latitude bands. Source: based on Bartlett and Harriss (1993)

is that this is more than offset by the loss of organic matter from these habitats when they are drained for agricultural or other use. The focus of this section is primarily on wetlands which actively accumulate organic matter. Although other wetlands may have large organic matter deposits, mires are more significant and accumulate the organic matter through a unique process. Mires actively accumulate or sequester organic matter, though they do so at a decreasing rate as they mature. The amount of organic matter from the anaerobic or waterlogged layer, which is emitted as CH_4 increases over time. The efficiency of the mire to accumulate carbon decreases as an equilibrium is reached. However, although savings in CH_4 emissions can be made if the mire is drained, the instantaneous release of organic matter more than offsets any emissions avoided.

Process of carbon accumulation in bogs

Peat-forming systems accumulate organic matter because conditions within them impede the decay of the plant material produced by their surface vegetation. This process is slow in terms of accumulation, in that bogs reach maturity over thousand-year timescales. Undisturbed bogs occur only where other human activities have not created alternative demand for the land. Bogs have been used for fuel sources and, where land was scarce and the market signals to land users created profitable alternative uses, drainage for agricultural purposes has occurred.

Bogs consist of two layers: an upper aerobic layer which allows water to drain through and is usually 10–50 cm deep; and a thicker waterlogged and hence anaerobic layer with a much lower rate of decay of organic matter. As the bog matures the peat mass tends towards a steady state in which the rate of addition of matter at the surface (through normal growth) is balanced by losses at all depths and the rate of accumulation is zero. These mature bogs therefore do not actively sequester carbon from the atmosphere, but rather act as a reservoir of stored carbon in much the same way as fossilised carbon strata, and are released either by human action (by drainage or extraction) or through natural changes in the ambient environment, such as climate and water table. Mature bogs can be in the order of 5–10 m thick over large areas or up to 70 m high in mounds. Indeed, carbon accumulation is almost unlimited given certain prevailing conditions, hence the assumption in Chapter 2 of 1200 t C ha^{-1} as a central estimate of storage of blanket bogs in the UK.

The water level defines the fuzzy distinction between the layers where the decay process is slowed by becoming anaerobic. According to Clymo (1984), the waterlogged part of the bog is the significant accumulating system, the rate being exogenously affected by the age of the bog (the decay factor) and the input into this layer from above, which is affected by the exogenous environmental variables of temperature, temperature flux and precipitation.

CH_4 and CO_2 trade-off

These variables also affect the rate of CH_4 production from bogs, which also takes place in the anaerobic section of the bog profile. If the accumulation of CO_2 is seen as a positive contribution to the carbon cycle in terms of reducing net emissions to the atmosphere, and the emission of CH_4 is perceived as a negative impact, then a trade-off exists between these activities. Indeed, the more productive the bog, in terms of plant growth, the greater the rate of CH_4 emission (Whiting and Chanton, 1993). Plants act as conduits for CH_4, siphoning CH_4 up through their roots and releasing it through their stems. The rate of CH_4 production also rises proportionately with the productivity of the wetland due to the high levels of leaf litter and below-ground organic matter in the anaerobic zone.

The draining of bogs reduces the anaerobic profile and hence reduces both the carbon accumulation and the CH_4 emissions. The area of bog has been decreasing through human intervention, primarily for agriculture and forestry activities. Globally, this disturbance may be in the order of 25 million ha in the period up to 1980 from the base area naturally occurring in the modern era (Armentano and Menges, 1986; Maltby and Immirzi, 1993). The remaining areas, across both temperate and tropical regions, sequester carbon at a rate possibly of 125 mt C year^{-1}.

However, as can be seen from Table 4.5, the impact of drainage for agriculture, forestry, fuel use and for horticulture globally, causes a large net emission annually higher than the sequestration rate of the existing bog. The changes of land use create a source of carbon but also bring about the loss of a sink. The global estimates reported in Table 4.5 are therefore the net impact of the annual loss of active mire and inactive peatbog. Further evidence of the net impact of draining and other exogenous changes to natural bogs are discussed in the

Table 4.5   Estimate of the loss of carbon from conversion of mires and bogs to other uses

	Former area (million ha)	Total lost (million ha)	Cumulated net emission as result of loss (million t C)	Annual loss (million t C)
Agriculture/forestry				
Temperate	399	20	4140–5600	63–85
Tropical	44	1.76–3.8	746	53–114
Fuel		} 5	590–780	32–39
Horticulture			>100	33
Total			5476–7126	181–271

Source: Based on Maltby and Immirzi (1993).
Notes: Net emissions reflect emissions from soils in subsequent land use or use of organic material as well as loss of annual active sequestration of bogs. Former area is estimated global area in 1795 (see also Armentano and Menges, 1986).

following section. The net impact of draining and loss of bog has also to consider this explicit greenhouse trade-off between CO_2, CH_4 and N_2O.

Carbon cycling in wetlands: possible future changes

Emissions from natural sources are strongly influenced by environmental variables such as soil temperature and inundation. These variables are likely to be affected by climate change, so the possibility of feedback has evoked considerable interest in examining the mechanisms which lead to changes in the fluxes described above.

The two major impacts of climate change are likely to be temperature and precipitation change, and wetlands are sensitive to both. Although no global analysis of the impacts of global warming and the possible feedbacks exists, there is now experimental evidence of some of the impacts. These are basically analogies of climate change impacts with other factors held constant. Freeman et al. (1993), for example, examine the effect of water table drop, analogous to precipitation loss in a Welsh peatbog, where Burke et al. (1990) examine possible global changes in wetland area due to temperature change.

The study of precipitation change is especially relevant for Northern wetlands as increasing summer droughts in Northern latitudes are part of some of the impact assessments. For the UK, mean precipitation may be higher in winter (6 to 15 percent) and of uncertain direction in the summer (-6 to $+13$ percent) by 2050, though the frequency of summer droughts will significantly increase (Hulme et al., 1993). The reduction in water availability through reduced precipitation and increased evapo-transpiration would threaten the factor which gives wetlands their unique properties. It has been suggested that the indirect effects of wetland hydrology are likely to have a greater impact on wetland methane and carbon dioxide production than temperature. Moore and Knowles (1989) and Freeman et al. (1993) both find reduced methane fluxes but increased CO_2 fluxes from reduced water availability (simulated through drawing down the water table). The results of Freeman et al. for drought of a Welsh bog *in situ*, are summarised in Figure 4.3. The mean fluxes of CO_2 from the control experiments rose by 146 percent (from 646 to 1590 mg CO_2 m^{-2} day^{-1}) at the peak of a simulated drought (shown as the dotted line and right-hand scale of the three graphs). Nitrous oxide fluxes also rose from a mean control level of 0.11 mg N_2O m^{-2} day^{-1} to a maximum 1.14 mg N_2O m^{-2} day^{-1} with the 20 cm fall in water table level. In contrast, methane fluxes fell by up to 80 percent (from 230 to 45 mg CH_4 m^{-2} day^{-1}).

These changes can be explained by the rates of metabolism that occur under anaerobic conditions which are lower than under aerobic conditions and hence increase CO_2 but reduce methane emissions. Increased aeration also causes

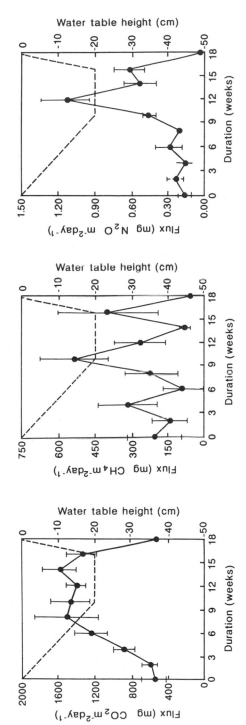

Figure 4.3 Carbon dioxide, methane and nitrous oxide fluxes from a Welsh peatbog after simulating precipitation decline from climate change. Reprinted from Freeman et al. (1993) with kind permission from Elsevier Science Ltd, The Boulevard, Langford Lane, Kidlington OX5 1GB, UK

nitrification, and since low nitrate concentrations limit nitrous oxide production through denitrification, the draining of wetlands allows greater nitrous oxide production. The net effect of these impacts depends on the relative changes in the fluxes and the radiative forcing potential of the gases, as discussed in Chapter 1. Although the results of Freeman et al. are site-specific and represent only one temperate wetland type, it is interesting to note the overall greenhouse gas feedback of the simulated precipitation change on its own. Given per volume global warming potentials for CO_2, CH_4 and N_2O in the ratios 1:11:270, the mean changes observed in the Welsh bog decrease the overall greenhouse impact immediately. The CO_2 equivalent emissions decrease by 25 percent from 3200 to 2400 mg CO_2-equivalent m^{-2} day^{-1}.

The feedback of temperature change on greenhouse gas emissions from wetlands is also likely to be significant, especially if it reduces the area of permanently frozen ground in Northern latitudes resulting in a greater area of anaerobic wetlands. The radiative properties of methane therefore contribute to warming and the warming enhances the production of methane through methanogenesis. The observed high concentrations of CH_4 of around 650 parts per billion by volume (ppbv) experienced in the interglacial period coincide with higher global mean temperatures and are often attributed to the large extent of tropical wetlands. Street-Perrott (1992) summarises the present controversy over whether tropical wetlands are responsible for this phenomenon. During past periods of high rainfall between glaciations, freshwater habitats expanded dramatically across west and east Africa, as observed from sedimentary records. Much of the enlarged wetland areas consisted of herbaceous swamps with high CH_4 emissions, thus explaining the high levels of atmospheric concentration. Further evidence comes from a short episode of cold dry climate at the end of the last glaciation, known as the Younger Dryas. At this time a sharp decrease in atmospheric methane of around 170 ppbv is observed from air trapped in ice cores, which points to the loss of freshwater wetlands as a significant factor in the overall atmospheric concentration.

In more recent history, Burke et al. (1990) estimate that methane emissions from all methane sources (primarily wetlands and landfill) have risen by 28 mt CH_4 between 1880 and 1980, representing a 34 percent increase. Using the same relationship between temperature and methanogenesis (but assuming that the area of wetlands will not change significantly by exogenous forces such as land use change), global temperature increases of 3 °C could bring about the resulting CH_4 emissions shown in Figure 4.4. Of course this static analysis does not take into account the precipitation changes or the other dynamics of land use change.

In summary, many of the future trends in greenhouse gas production from wetlands are highly uncertain with many of the forces pulling in opposite directions. Increased natural drought or the draining of wetlands for agricultural purposes reduces CH_4 emissions, but significantly enhances the conditions for CO_2 and N_2O production. Given historical land use trends, draining of wetlands

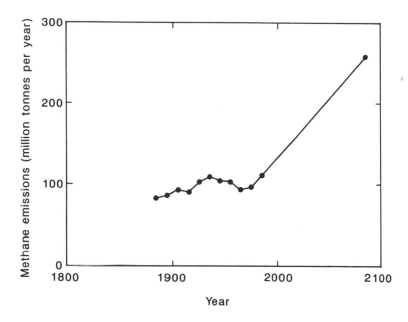

Figure 4.4   Estimates of global annual methane emissions from natural wetlands during the last century and up to 2080. Note: future values are calculated using estimates of 3 °C warming from a doubling of atmospheric CO_2 concentrations. Reproduced by permission from Burke et al. (1990)

will inevitably continue. The net effect of this will always be to result in carbon release (Heathwaite, 1993), though overall greenhouse forcing depends on the offsetting trace gas impacts. The net impact from the experiments carried out so far, suggests that a reduction of wetland area through draining or through future climate impacts on precipitation reduces immediate greenhouse gas net fluxes but increases atmospheric concentrations in the long run. As is a significant theme of this book, this is not a justification for draining, as global cycling is only one of the numerous ecological and economic functions of natural wetlands or other ecosystems.

   Global warming will undoubtedly lead to increased CH_4 production from existing natural wetlands, and may destabilise reserves of methane hydrates trapped in the permafrost in the Arctic. However, human-induced changes in the methane and carbon cycles of soil resources have been more important in recent centuries with significant new sources in landfilling waste, but also with a twelvefold increase in rice production in 100 years. Human intervention will influence future trends, as now discussed in the context of rice production.

4.6 RICE CULTIVATION

There are three main activities in agriculture which involve wetlands: paddy rice growing, the field processing of jute and similar coarse fibre crops, and certain types of aquaculture. Paddy rice growing is the most significant in terms of the area involved, the volume and value of the production, and the quantity of methane produced. This study therefore investigates the methane emissions associated principally with paddy rice production and, to a lesser degree, with coarse fibres; aquaculture is not included as it is not a significant source.

The estimates in this study of aggregate country fluxes are compared to the country-by-country estimates of previous studies and to other aggregate emissions estimates; they are found to be systematically higher. However, the purpose is not to allocate responsibility for the cumulative atmospheric concentrations which are causing the greenhouse effect. Rather it is to attempt to understand the processes that in future could contribute to the mitigation of emissions through management or through incentive policy, which would also be beneficial to rice farmers by increasing productivity. In short, the identification of the factors affecting methane emissions should be used to increase rice production per unit of input, while decreasing the methane flux per unit of rice production.

Paddy rice cultivation

Rice (*Oryza sativa* subspp. *japonica* and *indica*) is the only important annual food crop that can grow in waterlogged soils, the usual condition of many soils during the monsoon season in the tropics. Upland rice, grown on well-drained soils, only accounts for about 10 percent of the world's total rice area, and less than 3 percent of total rice production. Close to 150 million ha are used for paddy rice production in the world, most in the tropical and sub-tropical latitudes (Figure 4.1(b)). Of this rice area, 85 percent is in Asia (Figure 4.5).

Paddy rice land can either be rainfed or irrigated, the former being more common in south Asia, while irrigation is predominant in south-east and east Asia. Successful paddy rice growing is dependent on the supply and control of water to achieve moist soil conditions during the greater part of the growing period. A total of 1–1.5 m of water from rainfall or irrigation, at not less than 6 mm per day for modern paddy rice varieties, is generally needed (Wassmann et al., 1993). Usually, excess is applied to ensure that there is a gentle flow through the field. The rice seedlings are transplanted into a prepared well-flooded soil. As the plants grow, the depth of water is increased to 15–30 cm until flowering; from flowering to maturity the supply is gradually reduced until the field is almost dry for harvesting. During the growing period the field may be temporarily drained for short periods, especially for weeding and applying fertiliser. Depending on the variety and environmental conditions, the rice plant ordinarily matures between

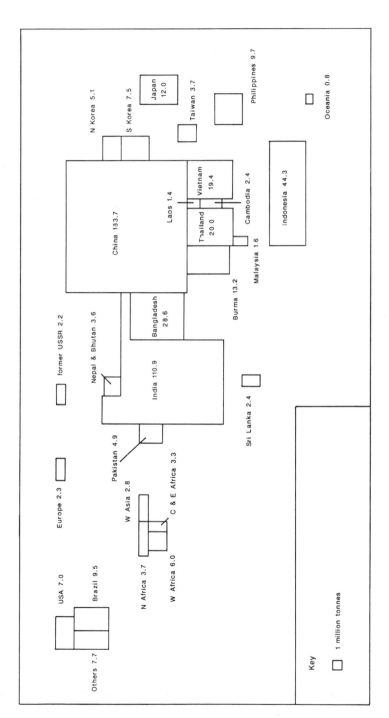

Figure 4.5 World rice production (1991) by country. Source: Mudge and Adger (1994)

about 90 and 200 days from sowing, with faster-maturing cultivars becoming more common (Purseglove, 1975). Methane emissions are then affected by the ambient environment which includes the soils and the fundamental process of anaerobic methane production, and by the intensity of cropping. These are discussed in turn.

Soils

An examination of maps of the distribution of tropical soils (FAO, 1991) and the main areas of rice production (FAO, 1992) shows that several different natural soil types are used for wet rice. However, because of the periods of submergence that the rice plant requires, these soils tend to acquire common characteristics, perhaps leading them to be treated as a distinct unit in spite of their differing geomorphological origins (Young, 1976).

Soils used for wet rice cultivation are characterised by a thick blue-grey reduced (gleyed) horizon, from 30 to 80 cm thick, overlain by a thin (2–20 mm thick) brownish oxidised horizon. If the soil was freely drained before being used for rice production, the reduced horizon will at depth pass into oxidised subsoil, with the boundary being marked by an impermeable pan, 5–20 cm thick, containing precipitated iron and manganese (Figure 4.6(a)). Originally poorly drained soils are gleyed throughout the subsoil (Figure 4.6(b)).

These distinctive profiles are produced by the saturation of the soil. In all soils there is a demand for oxygen, by the plants through their roots, and by micro-organisms. Most plants are killed by inadequate aeration of the soil, but rice is exceptional because it can grow without oxygen at the root level. The

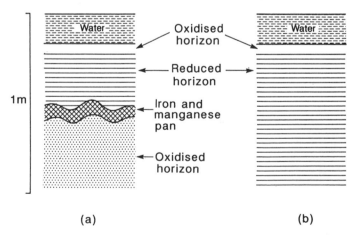

Figure 4.6   Paddy rice soil profiles (a) with reduced horizon caused by cultivation practices; (b) with natural reduced horizon. Adapted from Young (1976)

principal gaseous end products in the most severely reducing soils are a mixture of carbon dioxide and methane. The carbon dioxide content of the mixture of gases produced from rice soils varies from 100 percent in a high fertility soil where organic matter is metabolised exclusively by aerobic pathways, to just 7 percent in a low fertility, exclusively anaerobic soil (Watanabe, 1984).

Rice soils may be created either artificially by special cultivation techniques or they may exist naturally; in either case water control may be needed to maintain flooding. In the anaerobic soil conditions which exist below the surface layer of the soil, iron and manganese compounds, insoluble in their naturally oxidised state, are reduced to more soluble reduced compounds; natural leaching processes will carry these compounds in solution to lower depths. If free oxygen is present at these depths, as in an originally freely drained soil, these iron and manganese compounds will be re-oxidised and precipitated in the soil pores, forming an impermeable pan over a period of 50–100 years (Young, 1976). Such a pan will tend to reinforce the tendency of the soil to remain profoundly reducing (Figure 4.6(a)).

Even in soils which do not show an iron and manganese pan, an impermeable layer is deliberately induced by cultivation practices; pulverisation of the soil, repeated flooding with muddy water, puddling by repeated ploughings and treading, all tend to fill any soil pores and form a compact layer at the limit of cultivation, some 20–30 cm from the surface. This leads to waterlogging of the cultivated layer and, hence, reducing conditions (Figure 4.6(b)). Thus the soil conditions that are produced for wet rice growing promote the anaerobic decomposition of organic matter to methane.

In a flooded soil there is some dissolved oxygen in the surface water, but the demand within the soil greatly exceeds the rate of downward movement from the air through the water. Thus the surface of the soil is able to maintain its brownish oxidised condition while the underlying layers become reduced.

The schematic pathways of methane emission from a flooded rice field are shown schematically in Figure 4.7. Methane which is produced by the soil micro-organisms in the sub-surface layers of the soil is transmitted to the atmosphere by three routes, listed in order of importance:

(1) diffusion through the soil to the root zone, active transport by the plant aerenchyma to the leaves, and release to the atmosphere;
(2) diffusion through the soil to accumulate in bubbles sufficiently large to rise to the soil–water interface, followed by ebullition through the water into the atmosphere;
(3) diffusion through the soil and into the water lying above, followed by diffusion or convective transfer to the water–air interface, and diffusion into the atmosphere.

As the methane approaches the soil–water interface or the plant roots, it enters a region of oxidising conditions where aerobic micro-organisms convert some of the methane to carbon dioxide. This is either released via the water to the

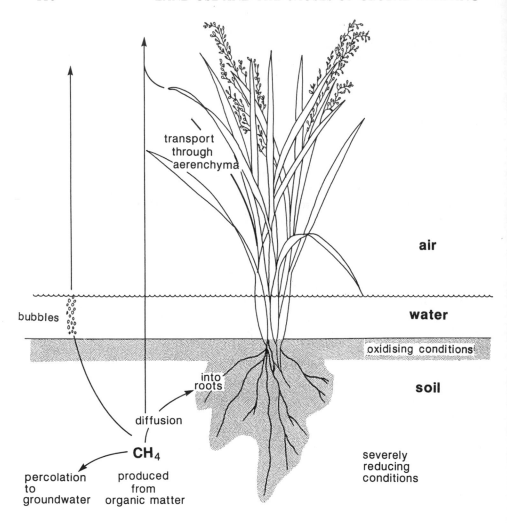

Figure 4.7   Methane generation, transport and emission in rice.
Source: Mudge and Adger (1994)

atmosphere or is used by the plant.

Tables 4.6 and 4.7 show the relative proportions of the different pathways as found experimentally during different growing periods. Methane formation in the soil depends on the quantity and type of organic matter in the reducing layer of the soil, on the soil micro-organisms, on the other soil constituents, on the temperature, and on the water management. The fate of the methane thereafter is determined by the balance between reoxidation and transport to the atmosphere. Reoxidation is mainly controlled by the water management and is strongly

Table 4.6 Methane oxidation in the soil and release to the atmosphere in different geographical locations

Country/growing period	Proportion of total methane oxidised in the surface layer of the soil (%)	Proportion of total methane released to the atmosphere (%)	Reference
Italy, early vegetation phase	> 50		Schütz et al. (1989)
Italy, late vegetation phase	up to 90		Schütz et al., (1989)
Italy, period not specified	80	20	Holzapfel-Pschorn and Seiler (1986)
Texas, whole vegetation period	58	42	Sass et al. (1991)

Sources: Wassmann et al. (1993), Bouwman (1990a).

Table 4.7 Methane transport routes to the atmosphere

Country/growing period	Proportional breakdown of the methane released to the atmosphere by transport route		
	Via plant aerenchyma	By bubbles in the water	By diffusion
Italy, early vegetation phase		Predominant, but falling with time	
Italy, late vegetation phase	Increasing with time, becoming predominant		
Italy, reproductive phase	> 90%		
Italy, whole growing season	c. 70%	c. 25%	c. 5%

Sources: Wassmann et al. (1993), Bouwman (1990a).

promoted by periodic draining of the field, allowing the surface to dry out and become oxidised by the air.

Methane transport to the atmosphere is subject to two influences: the soil and the plants. Low viscosity and a weak soil structure, determined by the nature of the soil and by the water management, allow bubbles to form and rise to the surface more easily, without significant reoxidation as they pass through the oxidised layer. The plant density of the crop and the rice cultivar will influence the methane flux, since more plants per square metre can transport more methane, and rice varieties differ in their propensity to transport gases through their aerenchyma.

Cropping

The rate of methane emission per unit area per year depends on the proportion of the year during which the soil is under anaerobic conditions. In many parts of the world only one annual rice harvest is possible, but irrigation, quicker-maturing varieties and more intense agronomic practices have enabled very many regions to grow at least two annual rice crops, and a considerable number even three, substituting for the traditional rotations of vegetables, jute, dry-land grain crops, oil-seeds, tobacco and potatoes. On a global basis, the approximate proportions are: 25 percent of the area single cropped with rice, 15 percent double cropped and 60 percent triple cropped. This increases the annual methane emission from each hectare and complicates the estimation of global emissions.

4.7   ESTIMATING GLOBAL EMISSIONS OF METHANE FROM RICE

The difficulties associated with converting knowledge of the processes into accurate values for global emissions are twofold: the portion of the year for which each hectare is under anaerobic conditions must be determined so that the total area can be found in terms of product of area and time; and the rate of methane emission must be estimated for each area–time unit. Previous methane emissions from paddy rice have presented estimates either by country or simply by latitudinal bands (Aselmann and Crutzen, 1989; Matthews et al., 1991; WRI, 1990, 1992; Subak et al., 1993).

The estimation of emissions by country is more relevant in the evaluation of past and present policies since international policy reviews have examined responsibilities by individual countries. Furthermore it is at the national level that influence on future emission strategies can be made, since land use is driven by interrelated institutional, demographic and economic factors. If least-cost strategies for reducing greenhouse gas emissions from paddy rice cultivation are to be promoted, this may involve agronomic practices which are less emission-intensive. This will clearly be a decision based on the interaction of national environmental and agricultural policies.

However, most estimations of methane emission at the country level are based on simplified methane fluxes and assumptions of area or production regimes. Bachelet and Neue (1993) compare the two methodologies:

(a) on a *production* basis:
    either
    Methane produced (g) = Rice produced (g) × Emission factor (g CH_4 per g rice)
    or
    Methane produced (g) = Organic matter incorporated into soil (g) × Emission factor (g CH_4 per g organic matter)
(b) on an *area* basis:
    Methane produced (g) = Area cropped (ha) × Portion of year under anaerobic conditions × Emission factor (g CH_4 ha^{-1} per time unit)

Whether the production or area basis is used however, the calculation of the emission factor is based on the same experimental data, usually reported on a g

CH_4 m^{-2} day^{-1} basis. The two alternative methods based on production seem to produce estimates which are some 20 percent lower than those based on area. The area method, which is used in this study, allows greater scope for refinement, as data on average rotation length and intensity of production can be incorporated directly. The approach adopted in this study to produce a revised global estimate is by area. It takes the most recent area and proportion of time under anaerobic conditions and refines the country-by-country estimates of methane per unit area to arrive at a revised estimate.

Estimating the artificial wetland area under anaerobic conditions

It is far from simple to derive values for the area which is under rice on a daily basis when conventional statistics give the area on an annual basis only. For example, FAO (1992) states that the 1991 world rice area was 148.4 million ha. Matthews et al. (1991) calculate that, taking into account the different cropping patterns, the 1985 annual world total of 147.5 million ha under rice (at some time in the year) is equivalent to 19 600 million hectare-days (the sum of all the areas under rice from day to day). They utilise data derived from earth observation imagery and from national statistics to estimate, to the nearest half-month, the crop calendars for each rice producing country. For China and India, the two principal producers of rice, their analysis is at the province/state level, rather than at the national level. In Matthews et al. (1991) these areas are converted into bands of 10° of latitude, giving the distribution shown in Figure 4.1(b). Bachelet and Neue (1993) use the same database to obtain, for the 18 principal producers in Asia, the monthly proportions of the total 1985 rice area used to grow rice in each country.

In this study the methodology is extended to all rice producing countries, using the most recent FAO figures for the areas on an annual basis, to obtain a month-by-month area. These areas are all converted into hectare-days by multiplying by 30. This area includes that planted with upland (dry) rice and deep-water (floating) rice, which do not produce significant methane emissions, and should therefore be excluded (WRI, 1992). The total number of hectare-days is therefore reduced by 17.4 percent to take into account these areas (Huke, 1982; WRI, 1992). This value for each country is the number of hectare-days which is multiplied by the methane flux per hectare-day to obtain the country's annual emissions.

Estimating the methane flux per hectare-day

Various estimates have been made for the methane flux from paddy fields. Matthews et al. (1991) and Bachelet and Neue (1993) apply the value of 0.5 g CH_4 m^{-2} day^{-1} to their estimate of the world rice area. On the basis of a recommendation by OECD experts, the World Resources Institute (1992) assumes that all paddy fields produce a uniform net flux of 0.44 g CH_4 m^{-2} per day of saturation. This contrasts with the value of 54 g CH_4 m^{-2} $year^{-1}$ (equivalent to 0.39 g CH_4 m^{-2} day^{-1} for a single 140-day rice crop) which WRI previously used in 1990. Subak et al. (1993) have also estimated the global emissions from paddy rice, using values for the daily methane flux derived from those given by Schütz et al. (1989) as 0.44 g CH_4 m^{-2} day^{-1} for temperate regions and 0.58 g CH_4 m^{-2} day^{-1} for tropical regions.

As discussed above, the methane flux is unlikely to be uniform, even within one farmer's holding, and certainly not on a regional, national or world basis, as soils, climate, rice cultivars and management all cause profound differences. This variability is confirmed by the relatively limited amount of experimental work that has been carried out. Ideally the flux should be determined experimentally for each soil/land use/management unit in each rice producing country, and these measurements should be updated as management practices change over time, but this is a counsel of unattainable perfection (Mudge and Adger, 1994).

Therefore the apparent precision of the flux values used in the previous estimations of global methane emissions given above is illusory. The true values for emissions per unit area may vary from 5 percent to 500 percent of the values used (Watson et al., 1992). One can merely hope that the average, over all the cropped areas during the whole growing season, is reasonably close to the values used.

Field measurements have been made in reasonable detail in five representative countries (China, India, Japan, Thailand and Italy) and show a wide range of values between countries and within countries as shown in Table 4.8. Adequate additional data is also available from California, also shown in Table 4.8, but this is not representative of current agronomic practices in the majority of paddy rice producing regions. For this study, two new methods of estimating each country's methane flux are used (and expanded upon in Mudge and Adger, 1994):

In *Method* 1, all paddy rice growing countries are classified into groups as being similar to one of six countries or regions for which an experimentally

Table 4.8   Methane flux values for selected sites used in Methods 1 and 2

Country/Region	Point flux value, Q (g CH_4 m^{-1} day^{-1}) for Method 1	Experimental extreme values of Q from literature	Values of Q obtained by regression (Method 2)
China	1.17	0.19 –1.44	1.16
India	0.40	0.01–0.66	0.42
Japan	0.24	0.11–0.39	0.25
Thailand	0.28	0.09–0.47	0.27
Italy	0.34	0.19–0.58	0.37
California, USA[*]	0.24	0.17–0.30	

Sources: Based on Watson et al. (1992), Khalil et al. (1991), Holzapfel-Pschorn and Seiler (1986), Wassmann et al. (1993), Yagi and Minami (1990). [*]California values not used in regression equation (see text)

Table 4.9   Grouping of countries for estimation of CH_4 flux per hectare of rice production (Method 1)

China	India	Japan	Thailand	Italy	California
$Q = 1.17$	$Q = 0.40$	$Q = 0.24$	$Q = 0.28$	$Q = 0.38$	$Q = 0.24$
North Korea	South Asia	South Korea	South-east Asia	Europe	USA
Egypt	Africa (except Egypt)	Taiwan	South America		Australia
	Oceania (except Australia)		Central America		
	Former USSR				

Note: Methane flux (Q) units g CH_4 m^{-2} day^{-1}.

derived flux is available. This classification is carried out by an appraisal on a macroagronomic scale of the corresponding climate, irrigation practices and level of technology. Then each country in the group is assigned the same flux value as the defining country or region (Table 4.9). Method 1 therefore produces a discontinuous range of flux values. The order of classification criteria is necessarily arbitrary: the most important factor is climate, followed by irrigation, followed by the level of technology. Thus the balance of the contributing factors in the grouping by country is subjective, and on casual inspection the countries appear to be grouped by climate.

Method 2 is a means of producing a continuous range of values, based on the observed experimental data. Different factors determine the emissions from the paddy rice field. As described in Method 1, the macroagronomic factors are very important, as they are expressed as the different soils' propensity to produce methane, but at the microagronomic level different contributory factors, such as the types of organic manure or different methods of land preparation, have not at present all been identified or ranked. Quantified data do not exist on a global scale, so other well-reported proxies must be found.

The nearest proxies for the various factors for which data exist in satisfactory detail are the yield of rice and the GDP per capita. The yield of rice per crop per hectare reflects the soil quality, climate, irrigation, fertiliser use (both organic and inorganic), and standard of management. An increased yield per hectare per year, from better soils and climate, more fertiliser or better management, an improved cultivar or an extra crop, implies greater quantities of organic matter, either as the rice plant itself or as organic fertiliser. These factors all tend to increase emissions.

GDP per capita is a loose proxy for technology used in rice production. Higher GDP implies the potential to use a higher level of technology, leading to more irrigation and more intensive management (and hence higher yields) but also a greater use of chemical rather than organic fertilisers (and hence less decomposable soil organic matter).

Thus, for a given level of GDP per capita and a given level of technology, higher yields tend to lead to higher emissions; for a given yield per crop, a higher level of technology implies a reduced use of organic fertiliser but possibly a greater number of crops per year.

Different combinations of these two proxies were regressed against the central value of the measured fluxes for the five countries where good quality data was available, and Box-Cox transformations were used to determine the best functional relationship between the flux and the explanatory variables. A regression equation explaining fluxes across countries then is:

$$Q = \exp{(-1.56 + 0.000341*YIELD - 0.0115*\sqrt{GDPpc}\,)}$$

$r^2 = 99.7$ percent,
r^2 (adjusted for degrees of freedom) = 99.4 percent;
significances: YIELD $p < 0.002$,
$\sqrt{GDPpc}$ $p < 0.002$
where

Q = methane flux (g CH_4 m^{-2} day^{-1}),

YIELD = rice yield (kg ha^{-1} for 1991) (FAO, 1992),

GDPpc = Gross Domestic Product per capita (US$ 1991) (World Bank, 1992). This single equation shows that methane emissions increase with increasing yield, but that increasing income in the country where the rice growing takes place tends to reduce the emissions per unit of production of rice. The 95 percent confidence limits are shown in Figure 4.8. The central value obtained by regression agrees very well with the experimental value. For China, the flux value is clearly much higher than for the remaining four countries; the confidence interval appears large, but even so, the lowest value is well above 0.8 g CH_4 m^{-2} day^{-1}. For the remaining countries the confidence intervals are narrower, and so the derived value is more certain.

For those countries where experimental work has been carried out, these regressed flux values are not judged to be significantly less accurate than the experimental values at a whole country level (as used for Method 1); in the case of those countries without experimentally derived values, they are certainly more accurate than either a subjectively derived value (as used for Method 1) or an 'across-the-board' value (as used in previous studies). As more experimental data become available, the range of values for use in the regression will increase, enabling a better relationship with narrower confidence intervals to be developed for those countries and regions within countries where experimental measurements of flux have not been made.

4.8   A NEW ESTIMATE OF GLOBAL METHANE EMISSIONS FROM PADDY RICE

The area-based methods discussed above for evaluating each country's methane produced from paddy fields give different results. All studies previous to the

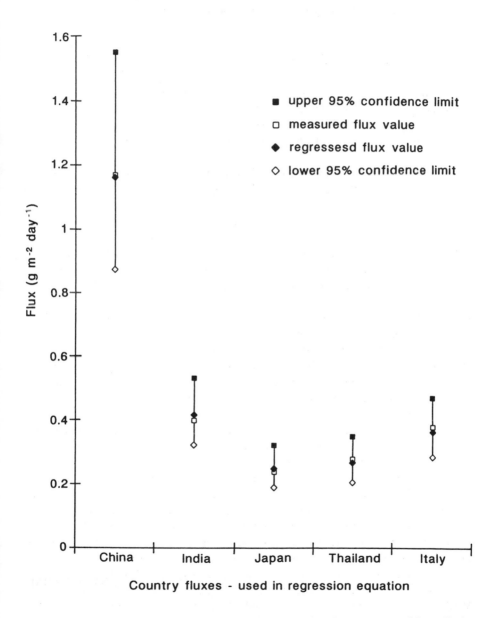

Figure 4.8   Methane fluxes—measured and regressed, with 95 percent confidence limits
of regression

present one state that they use the 'corrected' areas calculation of Matthews et al. (1991) area data supplied by FAO or the International Rice Research Institute (IRRI), but each uses a different value or values for the methane flux per unit area. These different results are presented in Table 4.10. The columns represent the following:

—Column 2: Subak et al. (1993) use 1988 areas and just two different flux values, depending on the region;
—Column 3: WRI (1992) uses 1989 areas and only one flux value;
—Column 4: the WRI method of column 3 is updated with more recent published area data (1991) and the development of the area analysis by Bachelet and Neue (1993), but still the same single flux value;
—Column 5: the latest area figures (1991) for each country (analysed by the Bachelet and Neue methodology) are used with a value for the daily flux for each individual country, derived by classifying countries into representative groups, one member of which has an experimentally derived flux value (Method 1);
—Column 6: the same areas for each country as obtained for Method 1 are used with a value for the daily flux for each country derived by regression from yield and GDP per capita (Method 2).

In Table 4.10 the area data therefore become progressively more recent, reading from left to right. WRI's 1992 global estimate of emissions (using 1989 areas) of 72.7 mt of methane (column 3) is revised slightly downwards when updated with 1991 areas, to 71.0 mt of methane (column 4). This may be because of the different method of deriving the areas between the two studies, since using the updated methodology with 1989 areas reduces the original WRI global estimate to 70.7 mt of methane.

The estimates in columns 4 to 6 vary solely with the different flux estimates used, as they all use the same estimates of area. The principal reason why Methods 1 and 2 of this study produce global estimates of 86.0 and 95.2 mt of methane, which are 21 percent and 34 percent higher than the updated estimate using WRI's methodology (71.0 mt of methane), is that the emission estimation for China is significantly different from previous studies. In both Methods 1 and 2 the estimate for China is over twice that of the updated WRI estimate.

In Method 1 most other countries' estimates are slightly lower than in the updated WRI estimate, but this is not apparent in the global total as China's greatly increased emissions, raised from 15.6 to 41.4 mt of methane per year, overwhelm the decreased emissions from the other countries. This arises because WRI uses a uniform flux estimate of $0.44 \text{ g CH}_4 \text{ m}^{-2} \text{ day}^{-1}$, whereas Method 1 assigns the flux value $1.17 \text{ g CH}_4 \text{ m}^{-2} \text{ day}^{-1}$ for China (the mean of the measured flux values from the main rice-growing areas of China), and flux values between

Table 4.10   Aggregate estimates by country of methane fluxes from paddy rice and coarse
fibre production

(1)	(2) Subak et al. (1993) (mt CH_4)	(3) WRI (1992) (mt CH_4)	(4) WRI (1992) updated (mt CH_4)	(5) This study (rice) Method 1 (mt CH_4)	(6) This study (rice) Method 2 (mt CH_4)	(7) This study (coarse fibres) (mt CH_4)
Total Asia	91.6	67.7	63.1	79.4	88.3	1.2
P. R. of China	25.0	18.6	15.6	41.4	41.1	0.2
India	26.0	19.0	19.4	17.7	18.3	0.5
Bangladesh	7.0	5.1	5.4	4.9	5.3	0.3
Indonesia	6.7	5.1	3.3	2.1	5.4	0
Myanmar	4.0	3.1	3.2	2.0	3.2	0
Japan	1.4	1.4	1.1	0.6	0.6	0
Taiwan	0.4	0.4	0.3	0.2	0.5	0
Thailand	8.2	5.7	5.3	3.4	3.3	0.1
Vietnam	4.4	3.2	3.3	2.1	3.8	0
Other Asia	8.4	6.1	6.2	5.1	6.8	0
Africa	2.9	2.3	3.8	3.8	3.4	0
North and Central America	1.3	1.0	0.9	0.5	0.8	0
South America	2.1	0.9	2.7	1.7	2.1	0
Europe	0.2	0.2	0.2	0.2	0.2	0
Oceania	0.1	0.1	0.1	0	0.1	0
Former USSR	0.4	0.4	0.3	0.2	0.3	0
World	98.4	72.7	71.0	86.0	95.2	1.2

Notes:
Column (2) Subak et al. (1993) use flux estimates of 0.44 g CH_4 m^{-2} day^{-1} for temperate and 0.58 g CH_4 m^{-2} day^{-1} for tropical rice areas; 1988 land use data.
Column (3) WRI (1992) uses 0.44 g CH_4 m^{-2} day^{-1} flux estimate; 1989 land use data.
Column (4) WRI(1992) updated same methodology but using FAO (1992) rice area data for 1991.
Column (5) This study (rice), Method 1 uses fluxes for countries in six groups, as explained in the text; FAO (1992) rice area data for 1991.
Column (6) This study (rice), Method 2 uses fluxes derived by regression, as explained in the text; FAO (1992) rice area data for 1991.
Column (7) This study (coarse fibres) emissions calculated from plant mass following a methodology adapted from Neue et al. (1990), assigned to countries in proportion to their fibre production in 1991 (FAO, 1992).
Totals may appear not to be exact due to rounding.

0.24 and 0.40 g CH_4 m^{-2} day^{-1} for other countries. Thus, for example, Indonesia's annual emissions decline from 3.3 mt of methane using the WRI flux, to 2.1 mt of methane when Indonesia is given the same flux as Thailand (0.28 g CH_4 m^{-2} day^{-1}).

In Method 2 the individual country fluxes are derived by regression, as explained above; China's flux estimate remains high (1.16 g CH_4 m^{-2} day^{-1}), and gives high annual emissions (41.1 mt of methane), while other countries have fluxes that more accurately reflect cross-country differences. Indonesia in Method 2 has an estimated flux of 0.70 g CH_4 m^{-2} day^{-1} based on its rice yield per hectare which is over 2.4 times that of Thailand and its potential propensity to use more organic manure and less mineral fertiliser than Thailand, while Thailand's estimated flux decreases to 2.7 g CH_4 m^{-2} day^{-1}; in this case Indonesia's estimated annual emissions more than double, and Thailand's decrease slightly.

Overall, the estimates from Methods 1 and 2 are significantly higher than most other estimates using regionally or globally averaged flux estimates, and tend to agree with that of Subak et al. (1993), who used flux estimates which this study would view as too low for high yield, low-income countries such as China and Indonesia, but too high for high-income countries such as Japan and Australia (Figure 4.9). In particular, this study gives estimates which agree better with those of the 1990 IPCC report and run counter to the downward revision from

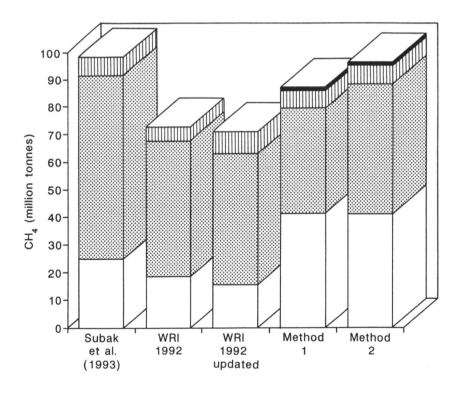

Figure 4.9   Comparison of global estimates of methane emissions

100 to 60 mt year^{-1}, advocated in IPCC's latest assessment (Watson et al., 1992).

The differences between the final global estimate (Method 2) and previous studies thus depend essentially on the flux value for China; if China's flux were to be similar to that of Japan, a country with a broadly similar climate and yield per hectare, then the emissions from China would be reduced by a factor of approximately 4.5, while global emissions would be reduced by a third. If, on the other hand, the experimental values from which this study derives its flux value are truly characteristic of paddy rice growing in China, and there is no reason to doubt this, then the particular agronomic practices which cause this constitute the primary target for mitigation strategies.

Since methane fluxes have usually been measured only during the growing season of a rice crop, and even then, not always continuously, it is likely that the final value may not take into account emissions such as those during wet fallows or from trapped bubbles of methane which are only released during disturbances such as weeding or subsequent cultivation and land preparation (Denier van der Gon et al., 1992; Neue, 1993). It is to be hoped that these underestimates are balanced by lower emissions that may arise because of events such as periodic draining for weeding. Continuous measurement over a period of several years, both during rice growing and when the land is otherwise used, is a pressing requirement for research.

4.9   OTHER ARTIFICIAL WETLANDS

The second most important agricultural practice that creates wetlands which produce methane is the initial field processing of certain coarse fibre plants. These plants include jute (*Corchorus capsularis* and *C. olitorius*), Kenaf (*Hibiscus cannabinus*), Roselle hemp (*Hibiscus sabdariffa*) and similar coarse fibre crops. World production of coarse fibres is approximately 3.7 mt, grown on 2.7 million ha (FAO, 1992), although other sources give higher values. For example Euroconsult (1989) estimates production at 4.2 mt in 1984. This higher value may correctly reflect the declining production of coarse vegetable fibres in the last 50 years. Most of the world's production is from Bengal (West Bengal in India, and Bangladesh). Other significant producers are China, Thailand, the former USSR and Brazil (FAO, 1992).

Two-thirds of coarse fibres are from jute, while of the remainder, the most important are kenaf and roselle hemp (Purseglove, 1974; Williams et al., 1980). Only these three are relevant to this study of anthropogenic methane production as the other crops do not involve the same anaerobic decomposition of organic matter. Roselle hemp is considered to behave as kenaf, as separate statistics are not available.

Methane is not produced during the growing of these plants but in the processing in the field to extract the raw fibres. The harvested plants are retted, a process in which they are steeped in ponds for periods of 14–21 days, until the aerenchyma tissues disintegrate in the water, allowing the fibrous tissues to be removed easily by hand. The canes which form the core of the stalk are dried for use as fuel, while the ponds are allowed to dry out so that the nutrient-rich mud can be used as fertiliser. It is estimated that 15 percent of the stem tissue decomposes in the retting ponds.

Neue et al. (1990) use production data to calculate the methane emissions from the anaerobic decomposition of the plant tissues in flooded paddy fields after harvest, a mixture of rice roots, rice straw, weeds and aquatic biomass. Although this methodology gives low estimates for the methane emissions from paddy rice, this study assumes that at least 12 percent of the anaerobically decomposing stem tissue in retting ponds is converted to methane, since the decomposing mixture in the flooded rice fields does not differ greatly from the decomposing tissue in the retting ponds. Table 4.11 shows the method of calculation and the consequent methane emissions. In Table 4.10 (column 7), these emissions are attributed to the different producing countries on the basis of their share of global production of jute and jute-like fibres in 1991 (FAO, 1992).

This additional quantity of methane, 1.2 mt, is small when compared to that produced by paddy rice, little more than 1 percent, but it does show that even comparatively minor agricultural activities have the potential to produce quantities of methane that merit inclusion in global estimates.

4.10 CONCLUSION

The emissions of methane from natural wetlands are in the order of 110 mt CH_4 year^{-1}, from the latest assessment by Bartlett and Harriss (1993). The methane flux is the largest natural *emission* of greenhouse gas that is completely independent of human intervention. However, human action continues to intervene in the system by altering the area of wetland, primarily by draining wetland for agricultural and other uses. As considered in Chapter 1, the causes of

Table 4.11 Methane emissions from coarse fibre plants

Coarse fibre crop	Production of dry fibre (million tonnes)	Fibre content of stems (%)	Production of stems (million tonnes)	Loss of organic matter during retting @15% of stems (million tonnes)	CH_4 emission @12% of organic matter (million tonnes)
Jute	2.4	5.0	48	7.2	0.86
Kenaf and Roselle Hemp	0.9	4.5	20	3.0	0.36
Total	3.3	—	68	10.2	1.22

Sources: Purseglove (1974), Williams et al. (1980), Euroconsult (1989).

this are the scale of economic development and policy incentives which encourage wetland loss. Although there are examples of conservation now being a higher priority and some of these policy distortions being reversed (Kramer and Shabman, 1993; Turner and Jones, 1991), the general trend continues. For example, the 'swampbuster' provisions in the 1985 US Food Security Act, which denied financial support to farmers if they undertook wetland drainage, where previously this had been subsidised. Despite this having a significant impact on the profitability of drainage, over 60 000 ha of wetlands were drained in the five years to 1987 (covering some post-reform years) in the Mississippi Delta states (Kramer and Shabman, 1993). Indeed field results seem to suggest that drainage will reduce the overall greenhouse gas flux when considering CO_2 and N_2O. However, the results are dependent on how much organic matter is lost, in other words on the CO_2 flux, and other assessments point to the net impact of drainage of wetland being to increase the atmospheric concentrations of greenhouse gases. In determining historical responsibility for the greenhouse effect, wetland drainage could theoretically be strictly accounted for in an historical assessment. But even if drainage could be shown to be beneficial in terms of greenhouse gases, this is not a justification for action, as wetlands are a unique ecological resource, as evidenced by the Ramsar Convention.

The other major human intervention, namely climate change itself, will undoubtedly increase the significance of wetlands. Fluxes historically have increased with increasing temperature, and the probability of release of methane from melting permafrost would be a significant climate change feedback.

The global emissions of methane from paddy rice growing presented in this chapter are approximately 95 mt CH_4 year^{-1} for current land use patterns. This is higher than previous estimates from country-by-country studies and higher than the latest IPCC 'best guess'. The reasons for other estimates being low are easily found: the studies take land use data from the 1980s, and use flux estimates which are generally based on a central low value. The estimates in this study use the latest land use data, and extrapolate the latest experimental flux data across all countries. Further experimental work will permit these estimates to be refined. The estimates of emissions from paddy rice can be extended to include other artificial wetlands if more agricultural activities are taken into account. The emissions from many seemingly minor activities other than rice production, particularly in the tropics, may be significant additional quantities of methane.

The future emissions of CH_4 from rice cultivation and other wetland agricultural practices are mainly dependent on the demand for these products, principally on rice as a staple food. Additionally the agronomic practices used in its cultivation will also determine the ratio of emission to production. A further uncertain factor is the potential feedbacks of climate change itself on the process. These issues are examined in more detail in Chapter 6, stressing that mitigation options are only likely to be taken up if they are consistent with the decisions taken at the farm level.

This chapter examines problems associated with defining and measuring rates of deforestation; sets out a framework distinguishing different causes of tropical deforestation; and subsequently summarises findings of recent studies which have attempted to test individual hypotheses for particular causes and regions. The development of policy measures to combat these causes is discussed and recommendations made.

5.2   DEFINING THE PROBLEM: DEFORESTATION AND FOREST DEGRADATION

Just as Chapter 3 highlighted difficulties in defining and measuring the areas of the different types of forest, so our knowledge of the exact rate of deforestation is hampered by the lack of a universal definition of *deforestation*. In the most complete surveys of the 1980s, deforestation was defined in several distinct ways: as the transformation of primary closed forest to any other formation; as the loss of any kind of closed forest; or as the lost of forestland, for example. No wonder that the range of estimates of forest cover and rates of deforestation is so variable.

Recent attempts by the FAO in the 1990 Forest Resources Assessment aim to overcome these difficulties and provide a more accurate evaluation of the current rate of deforestation. The definitions used in the 1990 Assessment are as follows: 'forests' are defined as ecosystems with a minimum of 10 percent crown cover of trees and/or bamboos, generally associated with wild flora, fauna and natural soil conditions and not subject to agricultural practices. 'Deforestation' refers to a change of land use with the depletion of tree crown cover to less than 10 percent. Changes within the forest class (for example, from closed to open forest) which negatively affect the site or stand, and in particular lower the productive (and thereby carbon sequestration) capacity are termed 'degradation'. Degradation is not reflected in the deforestation estimates (Singh, 1993).

So how do we delineate between deforestation and forest degradation, remembering that both may have profound effects on carbon dynamics? One approach is to examine the processes which are bringing about the transformation. Grainger (1993), for example, distinguishes between the two main human impacts on forests: deforestation and logging. These two are often confused, however, so following the FAO definition, Grainger refers to deforestation as 'the temporary or permanent clearance of forest for agriculture or other purposes'. So in this sense, forest has to be cleared and replaced by another land use to constitute 'deforestation'.

However, logging may also result in severe degradation of forest stands in a number of different ways. Adams (1992) makes a distinction between two types of logging: clear felling and selective logging. The intensity of selective logging varies, and Adams cites rates of removal of two to three stems per hectare in moist tropical forest in Africa, to more than 20 stems per hectare in parts of south-east Asia. Selective logging at high rates of intensity will have a number of different degrading effects which include the loss of nutrients and soil erosion, declining biodiversity both in terms of species mix and the genetic diversity of trees and the plant and animal species associated with them, loss of biomass, and changes in local climate including desiccation effects. In south-east Asia, Bruijneel (1992) reports that logging by tractor and skidder results in between 40 and 60 percent of all trees being killed or damaged (on top of those already harvested) and similar proportions of saplings are killed. In addition, 12–30 percent of soil surface is laid bare in the form of roads and skid tracks. In Queensland, prior to the introduction of strict silviculture practices in 1982, felling of only 10 percent of commercial species resulted in destruction or damage of more than 60 percent of the commercial tree population.

But harvesting methods are available which result in less damage. As Harrison (1992) points out, logging need not necessarily degrade the forest environment, provided that adequate time (many decades) is allowed for regeneration, and the least-damaging methods of felling and extracting felled timber are adopted. However, such benign extractive techniques require very careful management and, importantly, long term security of concessions, to provide incentives for sustainable utilisation (Flint, 1992). Such conditions apply in very few areas; in fact, in 1980, only 20 percent of logged forests in tropical areas were actively managed. Most of these were in Asia, where management usually meant little more than the existence of an official concession. In Africa, only 4 percent, and in Latin America 0.25 percent of forests are managed in any way (Harrison, 1992). Harvesting in these areas is more akin to mining. It is therefore likely that in many cases logging causes severe degradation of the ecosystem. However, this will not be captured in most deforestation statistics, and only the extraction of most of the trees (so that for instance in the case of the FAO Assessment, less than 10 percent cover remains) and clear felling as part of a logging operation, is considered. Again the data available will tend to underestimate the problem.

Given these caveats as to the measurement and definition of deforestation, the evidence as presented in Chapter 3 indicates that in many countries containing tropical forests exploitation continues at a rapid rate. Land conversion to other uses and unsustainable logging cause the loss of closed forests at a rate of nearly one percent per year. But what are the underlying causes of this process, and is it an inevitable part of the economic development of the countries concerned? Some authors (e.g. Grainger, 1993), believe that a certain level of deforestation is a necessary part of the development process for countries with large natural asset bases. The high levels of growth of economic activity (as measured through Gross Domestic Product) experienced by Indonesia in the 1970s and 1980s were based on the drawing down of the country's natural assets, primarily logging of natural forest and oil extraction (Repetto et al., 1989). The environmental and economic consequences of this exploitative development path are not wholly desirable, and although it is a commonly observed route, many would advocate development alternatives which do not exploit forest resources *in a destructive manner* (e.g.

Anderson, 1990). The following section identifies the explicit causes of deforestation in this economic development context.

5.3 THE CAUSES OF TROPICAL DEFORESTATION: DISTINGUISHING PROXIMATE AND UNDERLYING CAUSES

Deforestation concerns the change of use from forest land to some other use. Widespread deforestation has significant economic and ecological impacts, and there are large costs associated with excessive rates of loss, which include the costs of increased emissions of greenhouse gases, but also the loss of biological diversity, loss of other environmental functions such as watershed protection and soil conservation, and the costs borne by human beings who live and gain their livelihoods from forests (Panayotou and Ashton, 1993; Pearce and Brown, 1994). Although deforestation may provide some economic benefits, many of the costs are often not accounted for in making decisions about land use and conversion.

The proximate or direct causes of deforestation are the clearing of land for other uses such as agriculture, unsustainable exploitation of forest products, particularly timber, and natural causes associated with climate change and extreme events such as flooding or landslides and fires. However, there is also a set of forces which are applied at the local, national or international level which encourage or facilitate the unsustainable exploitation of forest resources (Pearce and Brown, 1994). These arise from two factors. The first is the competition between humans and non-humans for the remaining ecological 'niches' on land and in coastal regions; this competition reflects the rapidly expanding population growth, in developed countries in the past, and presently in the developing countries. The second set of factors constitute 'failures' in the workings of international and national economic systems such that these fail to reflect the true value of the environment. Essentially, many of the functions of tropical forests, for example the carbon sequestration capacity, are not marketed, and as such are ignored in decision-making. Additionally, decisions to convert tropical forest are encouraged by fiscal and other incentives.

These different causes are discussed by a number of authors (e.g. Swanson, 1994) and different terminology is used to describe them. For example, Panayotou and Sungsuwan (1994) make the distinction between activities such as logging, fuelwood collection and land clearing for agriculture as *sources* of, as distinct from *causes* of deforestation. Flint (1992) discusses the causes of biodiversity loss (of which loss of tropical forests is a major component) and distinguishes between *immediate causes* such as habitat loss and overexploitation on the one hand, and *underlying causes*, which are less clear cut and easily identifiable, on the other. Flint divides these underlying causes into three groups: development pressure, market failure and intervention failure. Development pressure is characterised by Flint as increases in population, exacerbated by the unequal distribution of agricultural land, and by the replacement of traditional common property regimes by open access regimes. Market failures occur for five reasons: ill-defined property rights, externalities, uncertainty and irreversibility, market imperfections and policy distortions. Intervention failures result from either ineffective positive intervention, or from unintentional negative intervention. These in turn occur either as part of the general development strategy: in fiscal policy; monetary policy; market interventions; or land tenure polices. Table 5.1 presents a simplified framework of these proximate and underlying causes of tropical deforestation, showing that observation of the conversion process will show what the direct causes are, but a complex set of circumstances, some defined in remote capital cities or even internationally, contribute to deforestation.

We will now examine each of the different factors propounded as causes of tropical deforestation and the effective policy measures needed to combat their effects. We exclude the natural causes of deforestation shown in Table 5.1 for the present as, although there are arguably policies which can be followed in order to mitigate these effects, they are generally outside human control.

5.4 PROXIMATE CAUSES OF DEFORESTATION

This set of causes is the most easy to identify and recognise, and consists of a direct causal agent, action or process. So this includes the conversion of land to other uses such as subsistence or commercial cultivation, cattle pasture, and developments such as dams and mining. It also includes exploitation by timber companies, and a number of authors and environmental pressure groups (for example, Friends of the Earth in their campaign against the trade in certain tropical timber species) have levelled blame for high rates of deforestation in some tropical countries at the international timber trade. These proximate causes, as set out in Table 5.1, are each discussed in turn.

Table 5.1   The causes of tropical deforestation

Proximate or direct causes	Underlying or indirect causes
Land use conversion to:	Population increase
Subsistence agriculture	Poverty
Cash crop/ plantation agriculture	International economics:
Cattle ranching	Debt and macroeconomic adjustment
Other developments mining, dams, etc.	Policy failures:
Overexploitation of forests:	Roads
Timber	Subsidies for land use conversion
Fuelwood	Migration and colonisation
Natural/environmental:	Market failures:
Climate change	Failure to capture 'public good' aspects
Floods, landslides	of forests

Land use change

Most of the competition for space between human and other species is demonstrated by the conversion of land to agriculture, infrastructure, urban development, industry and unsustainable forestry. Data presented in Chapter 1 showed historic changes in land use, and Table 5.2 gives a generalised picture of land use conversions by world region between 1977 and 1987. This table implies that in North, South and Central America much of the forest area has been converted to pastureland during this period, whereas in Asia and Africa, most change is associated with other sorts of development, including roads, urbanisation and, we assume, degraded forest.

The depletion of natural habitats and the expansion of agriculture concurs with the classic Malthusian thesis predicting agricultural extensification with increasing human population. Malthus, in 1798, postulated a tendency for human populations to grow geometrically, but for the means of subsistence to grow only arithmetically. The former will outstrip the latter over time, resulting in a 'Malthusian crisis'. Globally there is little to suggest that this thesis has been proven because although the extensification of agricultural land use continues, at the same time intensification of production (as postulated by Boserup, 1965) and out-migration both occur and the exact relation of these processes to environmental degradation, and especially deforestation, is still the subject of intense debate (see Section 5.6 below) and is dependent on contributory factors such as property rights, particularly whether open access conditions operate (Swanson, 1994; Pearce and Brown, 1994). What is clear from Table 5.2, however, is that Africa, South America and Asia are undertaking net conversion from forest to other uses, whereas North and Central America, and Europe have experienced a net gain in forest area. This is illustrative of what Mather (1990) has termed the 'Forest Transition', which postulates that the level of economic development and population density determine the demand for forest products and thus forest cover. So what processes and agents are

Table 5.2   Global land conversions 1977–1987

	Cropland (million ha)	Pasture (million ha)	Forest (million ha)	Other (million ha)
Africa	+8	−4	−25	+22
North and Central America	+3	+11	+7	−20
South America	+14	+20	−41	+11
Asia	+4	−2	−29	+29
Europe	−2	−3	+2	+4

Source: Pearce and Brown (1994).
Note: 'Other' land includes roads, uncultivated land, wetlands, built-on land.

causing the conversions of land use? The following sections briefly examine a number of factors: subsistence agriculture, cash cropping, cattle ranching, fuelwood, mining and other infrastructural and industrial development, and the trade in tropical timber.

Subsistence agriculture

Myers (1994) lays blame for the largest share of tropical deforestation squarely on what he refers to as the 'shifted cultivator' which he claims accounts for 61 percent of total deforestation. He describes how, until the 1960s, shifting cultivation as practised throughout the tropics, was a sustainable method of land use. Areas of forest were traditionally cleared, cultivated for a few years and then left to revert to forest. Fallow periods were sufficiently long to ensure regeneration of vegetation and soil nutrients, and land use was extensive enough to avoid any widespread degradation of the environment, e.g. significant losses of biodiversity or watershed disruption. As long as there was plenty of available land, this constituted sustainable use of forests. However, in the last 25 years, increasing numbers of landless and otherwise displaced peasants have made more severe demands on forest areas. Many of these farmers migrate large distances in search of land to fulfil their need for subsistence, and they have often not been able to develop cultivation systems appropriate to the forest ecosystem. The shortened fallow periods and inappropriate methods do not allow for forest regeneration. These different systems of shifting cultivation are characterised by Grainger (1993) as either *traditional shifting cultivation*, which is a proven form of sustainable land use in the humid tropics; *short-rotation shifting cultivation*; or *encroaching cultivation*. These two latter forms are unsustainable in the long run. Grainger (1993) estimates that shifting cultivators comprise the majority of forest-dwellers, numbering 200–300 million people, whereas Myers (1994) estimates 300–500 million 'shifted cultivators' worldwide. Likewise, the area of forest affected differs, with estimates ranging from 7 to 20 million hectares each year (National Research Council, 1993).

Although small farmers and shifting cultivators may be the main agents of land use change, any exploration of the forces driving this change reveals a complex set of influencing factors. For example, Colchester and Lohmann (1993) argue that the problem starts with the perverse and inequitable agricultural policies operating in many developing countries. The erosion of land rights in agricultural systems results in the marginalisation and impoverishment of small farmers and leads to the colonisation of forest areas, which is further fuelled by government policy and the development of roads and other infrastructure. Poverty and lack of secure land tenure therefore lie at the root of the problem. Some of these policy issues are discussed in the next section on the underlying causes.

Cash crops and plantation agriculture

The expansion of cash cropping and plantation agriculture has, in some parts of the world, contributed to deforestation. This is particularly true in certain areas; for example, the development of sugar cane plantations has been a major cause of forest loss in the Caribbean. It dates from the start of European colonisation in the 16th century, but is continuing on some islands to this day. Of the ten leading primary commodities exported from developing countries, four are tree crops which are often grown on land converted from forest. These are coffee, rubber, cocoa, and vegetable oils derived from oil palm and coconut palm. The sustainability and long term impacts of their cultivation on greenhouse gas emissions will depend both on their system of management, as well as on the previous land use. It is possible that the greenhouse gas fluxes associated with their cultivation may result in lower net emissions, or higher net sequestration, than climax forest. Rubber and oil palm plantations can produce stable or increasing yields on a long term basis; for example, rubber has been grown for over 100 years, and oil palms for more than 70 years on some sites in Peninsular Malaysia (National Research Council, 1993). However, maintained and increasing yields are achieved through the application of artificial fertilisers and pesticides, so the sustainability of the systems is rather dubious. Other crops, such as sugar cane, have greenhouse gas emissions associated with their cultivation through the use of burning in the production cycle.

Cattle ranching and the 'hamburger connection'

Table 5.2 shows that conversion of forest to pasture is significant in Latin America, but not in either Africa or Asia. Cattle ranching as a cause of tropical deforestation has achieved notoriety through critical accounts of the 'hamburger connection' (Myers, 1981), linking forest clearance in Central America with North American fast-food outlets through beef exports. During the 1970s, US beef prices rose, so that by 1978 Central American beef was less than half the wholesale price of US beef. As a result US imports increased, and in 1978 Costa Rica exported five times, and Guatemala 15 times the amount of beef exported in 1961. This opening up of such a large and accessible market in North America, the argument goes, provided an incentive to clear large areas of forest for conversion to cattle pasture. Conversion was further encouraged by government policies offering export, tax and credit concessions. Although intuitively the direct link between hamburger consumption and deforestation is appealing, and illustrates how the globalisation of the world economy and the remoteness of production and consumption contribute to external environmental impacts, the link is not accepted as the sole mechanism of pasture-led deforestation. Other distortionary policies enacted by governments in the countries of Latin America ensure high financial returns to cattle ranching as a form of land speculation

(Hecht, 1993). Further, the regulatory situation in many countries encourages forest clearance and ranching as a method of securing title deeds. A combination of factors promote cattle ranching at the expense of standing forests, and at the expense of inequitable development.

The conversion of forest to pasture has long term ecological and environmental effects; although forest clearance brings about the release of soil nutrients this effect is only short-lived, and productivity of pasture rapidly declines. This means that ranches must continue to expand and clear new areas of pasture simply to maintain their herd size. Indeed in some regions, ranches may be abandoned after as few as five years. The expansion of cattle ranches has in many places had seriously disruptive effects on smaller producers and people dependent on extraction of forest products (for example, rubber tappers in Brazil) and on indigenous groups of forest- dwellers. Cattle ranching can therefore be seen to be socially regressive, as well as environmentally damaging (Adams, 1992).

Fuelwood

The use of wood for fuel is one of the principal extractive uses of forests. The greenhouse gas implications of this use and of improved technologies for using this, depend on whether this action is part of a closed sustainable process, or whether it is a precursor to deforestation. In developing countries up to 90 percent of domestic fuel needs are met by woodfuel (wood and charcoal), and woodfuel may account for 80 percent of the wood harvested. However, only a small proportion of wood extracted from moist tropical forests is actually used by local people as fuel, and the collection of woodfuel tends to cause more destruction in drier forest areas, and close to large population centres. Eckholm et al. (1984) discuss the role of woodfuel collection as a cause of deforestation, and conclude that this is only a problem around large cities. Bennett (1992) estimates that cutting fuelwood may be responsible for up to 10 percent of forest destruction, but that most of this is at the forest edge. In general, users prefer to collect deadwood, which is drier and therefore easier to carry and to burn, rather than cutting down trees. However, as wood becomes more scarce people may be forced to cut trees; as long ago as 1980, FAO estimated that up to 1.3 billion people, or 39 percent of the population of developing countries, live in woodfuel-deficit areas.

Collection of fuelwood for rural use is generally carried out on a small scale, and deadwood and twigs are collected primarily from trees and bushes on farms and other non-forest sources. According to Mather (1990), the proximity of source to users is a significant factor in firewood collection of this type not causing deforestation. Collection for urban use, however, is likely to take place on a much larger scale and where a market exists, there will be an incentive to cut live trees, and collectors will venture into forests. In addition to domestic use, there are also increasing demands for fuelwood for use in cottage industries and rural processing activities. These include agricultural processing activities such as tobacco-curing and tea-drying, which are estimated to account for 11–25 percent of fuelwood use in developing countries. These demands may cause severe deforestation around production centres; for example, it has been calculated that 1 ha of tobacco requires fuelwood from 1 ha of savanna for its processing.

As woodfuel becomes increasingly scarce and a valuable, traded commodity rather than a freely available good, exploitation of forests for profit ensues. This is evident in recent years in Kenya, where illegal charcoal production and sale provides income for many people in a depressed economy. Some forested countries are also exporting charcoal; for example, Thailand and Indonesia export charcoal to the oil-rich Gulf states, where it is a preferred cooking-fuel (Brown, 1985). The conversion of wood to charcoal, often undertaken on a small scale with inefficient technology, results in conversion losses of 2.5:1 on average. Two and a half times as much wood is therefore used by a household using charcoal as one using wood as fuel (Mather, 1990).

So although the demand for woodfuel may be a primary cause of deforestation to a localised extent, e.g. around urban areas or agro-processing centres, it is not generally a major cause of large-scale deforestation of moist tropical or closed forests. If no net loss of biomass ensues from fuelwood use, the only greenhouse gas implications are related to nitrous oxide and CH_4 emissions in the burning process.

Mining, dams and other uses

In recent years much media attention has focused on the destruction of tropical forests to make way for developments such as mining and other industrial operations, and dams and hydro-electric plants. Whilst such developments may not have a particularly significant overall impact on total global deforestation, their impact locally may be extremely deleterious. Not only do they cause deforestation in their immediate vicinity but there are often high levels of pollution and other destruction associated with the roads and other infrastructure accompanying developments. In addition, these developments may cause the migration of large numbers of people into the area, to work on the construction and operation, and who clear land for settlement and agriculture. This may also lead to disputes over property and land rights, and a number of cases (for example, the plight of the Yanomani Indians in Brazil and immigrant gold miners) highlight problems associated with migration to remote forest areas and conflicts with indigenous people and their use of and rights to forest resources. These developments, although destructive enough in their own right, make their biggest impact on deforestation by opening up areas for settlement and colonisation as well as having downstream effects which may extend to the regional environment.

International tropical timber trade

Extensive logging of tropical rainforests is a phenomenon of the last half century. In this period tropical hardwood exports have risen 14-fold, from 4.6 to 61.2 million m^3 roundwood equivalent per annum between 1950 and 1980 (Grainger, 1993). The demand was led initially by European countries, but in more recent years, Japan has become the world's leading importer of tropical hardwood. This rise in international trade has resulted in a 'boom-and-bust' export pattern where high initial export earnings were followed by the depletion of old-growth forests and the collapse of domestic processing industries (Vincent, 1992). This pattern emerged in West Africa during the 1950s and 1960s, and became even more apparent in the 1970s and 1980s as trade shifted towards south-east Asia and expanded in volume.

The three factors most often blamed for this boom-and-bust, inherently unsustainable pattern of timber exploitation are: first, that consumption in developed countries drives the boom; secondly, that import barriers inhibit the development of processing, and therefore the value of forests to tropical countries; and thirdly, that low international prices represent market manipulation by the developed countries which reduce the financial viability of forest management strategies. Vincent (1992) dismisses these three popularly-held misconceptions of the role of the international timber trade in causing tropical deforestation. In the first instance, only about one-third of the industrial roundwood harvested in developing countries enters the international market in any form, and as global consumption of wood products rises (although at a diminishing rate), developing countries account for an increasing share of the market, so that demand for tropical hardwood is divided roughly equally between temperate and tropical countries (Grainger, 1993). Vincent contends that developed country tariffs on wood products are generally lower than corresponding import tariffs in developing countries, and that in many cases, are lower than developing countries' export taxes on wood products. Concerning low prices, it appears that competition with temperate timbers has inhibited increases in the international prices of tropical timber products, and the prices for tropical hardwood products tend to fall between those of temperate softwoods and hardwoods. In the last 50 years the average export prices for tropical hardwood logs and sawnwood have been substantially lower than corresponding prices for temperate hardwood products. As roundwood supplies from temperate countries increase, it seems unlikely that there will be any substantial increase in tropical timber prices, so this mechanism, for the time being, will not offer an incentive to more sustainable management and exploitation of tropical forests for timber.

Of course, selective logging, the predominant form of timber exploitation, does not necessarily lead to deforestation or even degradation of forests. Earlier sections of this chapter described how selective logging often destroys large numbers of trees and may lead to soil erosion and other forms of pollution.

Selective logging should not be a haphazard affair, but a system of management relying on careful extraction followed by natural regeneration for future harvests. Possibilities for implementing more sustainable methods of logging are discussed in Section 5.6, but it is clear that the present system of international trade offers no incentives for sustainable extraction, encouraging instead the boom-and-bust system. This is further encouraged by the systems of property rights and concessions which control logging in most tropical countries, and exacerbated by a number of policy failures which are discussed in Section 5.5. Harrison (1992) maintains that logging's biggest role in deforestation is as a facilitator for shifting cultivation, whereby logging roads provide access into the forests by settlers. The role of logging in opening up areas for agricultural colonisation is difficult to quantify, but is undoubtedly a very important one.

It is not necessarily the international trade which causes deforestation, but concessions policies contribute towards the unsustainable exploitation of forests. Panayotou and Ashton (1993) claim that the most serious problem faced by the tropical timber trade is the undervaluation of the resource compounded by five factors to do with the way in which timber concessions are administered:

(1) insecurity of tenure resulting from logging concessions that are shorter than felling cycles, lack of concession transferability, and uncertain renewability;
(2) logging concessions that are awarded on a political rather than an economically competitive basis;
(3) regulations that require concessionaires to begin harvesting their sites by a stipulated time;
(4) tax structures based on marketable timber removed, rather than the potentially marketable timber on site, thereby encouraging high-grading and damage to the remaining stand; and
(5) the disregard of customary use rights, which leads to interference and encroachment on concessions by members of local communities.

The policy measures available to improve concessions policies so that they provide incentives for more sustainable management are discussed later in this chapter.

5.5   THE UNDERLYING CAUSES OF DEFORESTATION

The indirect and underlying causes of deforestation are far more complex and controversial than the direct causes discussed above. Authors have variously postulated the rise in human population, macro-economic factors such as indebtedness, international trade and exchange rates, or government policies and micro-economic factors as being to blame for excessive tropical deforestation. Many studies have examined the impacts of particular factors on the exploitation of forests in individual countries. However, when the interactions between different factors are tested across a range of countries to try to identify causal relationships, the picture becomes far less clear. Again we will first examine the issues individually.

Population

Increases in the human population of developing countries is most often cited as the major underlying cause of deforestation. Empirical evidence highlights the complexity of factors concerning population pressure and its links with agricultural extensification and intensification. These would appear to imply that whilst the initial response to increases in population pressure may be agricultural *extensification*, when land becomes more difficult to access or only very marginal areas remain, then *intensification* occurs. These two processes are at the core of the Malthusian and Boserup explanations of the land use response to increases in human populations. However, empirical evidence suggests that both processes happen, perhaps even simultaneously, and it is not fully understood what factors determine which response in what particular circumstances. For example, if the global figures on the yield and area of cereals are examined in relationship to population, then both processes are evident. In those parts of the world that still have large reserves of land and low population density, a big share of the increase in food production since 1961 was due to area expansion—no less than 51 percent in Africa. But in regions with the smallest land reserves and highest population densities most of the increased production came from yield increases. Yield increases accounted for 79 percent of growth in cereal production in Asia in the last three decades. Between 1983 and 1987, yield increases were responsible for all production increases (Harrison, 1992). Growth in cereal area tends to be fastest in areas of fastest population growth; growth in yields, in contrast, is slowest where population growth is fastest, and fastest where population growth is slowest—the reverse of the Boserup thesis. This implies that population growth as such does not stimulate yield increase until land shortages begin to develop; in other words, accessible forest is likely to be cleared first.

Evidence also suggests, however, that the next phase in this progression, is reforestation, what Mather refers to as the forest transition (Mather 1990), and Harrison points to recent history in northern Europe (for example, in the UK forest cover has increased from 5 to 10 percent in the last 50 years) and China. Recent research by Tiffen et al. (1994) indicates that this process is also underway in certain parts of Kenya, a relatively densely populated developing country. Their study examined historical land use changes in Machakos District, an area of high population growth. This revealed that during the first half of this century rapid deforestation—land clearing for agriculture—took place. The trend continued until the 1960s; between 1948 and 1961 forest area declined by 21 percent, but between 1961 and 1978 it reversed and forest cover increased by 165 percent. This is as a result of individual tree planting on farms and community forestry programmes; as trees and their products become more valuable, then there is an incentive to plant trees on private land. This case is not necessarily part of a natural progression, and it is assumed that certain conditions are required for trends in environmental degradation, and especially deforestation, to be reversed. These are likely to include a private and secure system of land tenure and access to markets and information, which facilitate technological innovation (Tiffen et al., 1994). However, this case does illustrate the fact that long term decline in environmental quality and forest cover is not necessarily either inevitable or irreversible in developing countries experiencing high rates of population increase.

In general then, the evidence linking overall increases in the human population and the rate of deforestation is mixed, and although intuitively it would seem reasonable to expect that more people require more food and, in turn, more land which results in less forest, the above discussion shows that this is not always the case. A number of studies have tested various different indicators of population (e.g. growth rates and rural population density) and the results across countries are by no means conclusive. Some of these studies are summarised in Table 5.5. Population pressure is one factor in the deforestation equation, but its effects will be different in different circumstances and in response to other factors. For example, Repetto and Holmes (1983) argue that population growth together with open access, asymmetric tenure and commercialisation with increasing international demands, leads to considerably faster, even accelerating deforestation than population growth alone. Yet again, policy and economic factors are important in deciding the extent of deforestation.

Economic, market and intervention failures

Conventional economic approaches to the valuation of forests fail to account for the non-timber forest products and services in forest management and investment decisions. In many cases, the only product of tropical forests which is considered of economic value is the timber produced. However, a whole range of non-timber forest products, including fruits, latex and fibres, as well as the environmental and ecological services and functions, such as soil protection, water cycling and carbon storage, are not valued. Environmental economics attempts to overcome this by considering total economic value. This is a concept which aims to cover the range of different uses and services associated with an environmental asset. The range of different aspects of 'total economic value' of tropical forests is shown in Table 5.3.

The total economic value (TEV) can be expressed as:

TEV = Direct-use value + Indirect-use value + Option value + Existence value

Direct-use values include timber values and revenue from non-timber forest products. *Indirect-use values* which may be referred to as functional values, relate

Table 5.3   Total economic value of a tropical forest

Use value			Non-use value
(1) Direct value	(2) + Indirect value	(3) + Option value	(4) + Existence value
Sustainable timber Recreation Medicine Non-timber products Plant genetics Education Human habitat	Nutrient cycling Micro-climate Air pollution reduction Watershed protection Carbon fixing	Future uses of (1) and (2)	Forests as objects of inherent value, as bequest as gift, as responsibility

Source: Pearce (1991a).

to the ecological functions performed by forests, such as global biogeochemical cycling, the protection of soils, and the regulation of watersheds.

Option values relate to the amount that individuals are willing to pay to conserve forests for future use. Option value is therefore like an insurance premium to ensure the supply of an asset, the availability of which would otherwise be uncertain. Quasi-option value assumes that the expected value of an asset will change over time due to an increase in the stock of knowledge. The classic quasi-option value is that of genetic resources, where forest conservation incurs the quasi-option value of future discoveries of genetic resources such as for pharmaceuticals.

The *existence value* relates to valuations of the environmental asset unrelated either to current or optional uses. Many people reveal their willingness to pay for the existence of environmental assets through wildlife and other charities without ever partaking of consumptive or non-consumptive uses. Empirical measures of existence values through methods such as the contingent valuation approach suggest these can be a significant element in total economic value.

The total economic value does not claim to capture all aspects of value. Any human appreciation of intrinsic value can only be assumed to reside within existence values. Although some environmental economists might claim that it does represent all the economic value, many ecologists believe that TEV does not account for the whole economic story. There are still some underlying functions (termed 'primary values') that are prior to the ecological functions and services and are essential system characteristics upon which all ecological functions are contingent. There are also dangers in assuming that it is possible to calculate TEV by aggregating the components: some values may be mutually exclusive in that the use of an asset for one function, precludes or lessens its value in another aspect. In other words, there may be trade-offs between the different components of TEV. For example, maximising the direct use of forests for timber may lessen the values for recreation or biodiversity.

Pearce and Brown (1994) distinguish three types of economic failures that result in high rates of deforestation in tropical countries: local market failure, global appropriation failure, and intervention failure. Local market failure is the classic economic cause of under-investment, where market forces are not able to secure the economically correct balance of land conversion and forest conservation. An underlying assumption, of course, is that there is an economically optimum rate of deforestation, which is not zero. Local market failure arises because those who convert the land do not have to compensate those who suffer the local consequences of that conversion; extra pollution and sedimentation of waters caused by deforestation, for example. The corrective solutions to this problem are well known: a tax on land conversion, zoning to restrict detrimental land uses, environmental standards, and so on.

The rate of return to forest conservation is distorted by what economists call 'missing markets'. What this means in the tropical forest context is that systems of habitat and species are serving valuable functions which are not marketed. Effectively, then, no-one values these functions because there is no obvious mechanism for capturing the values. Local market failure describes this phenomenon within the context of the country or local area. But there are missing global markets as well. This can be illustrated by the example of the value of carbon storage by tropical forests. Other global benefits of tropical forests, such as non-use and existence values are also not reflected in global markets. Such values can be estimated using a number of different techniques including contingent valuation (CVM), travel cost method, hedonic property price and production function approaches. Global appropriation failure arises because these values are not easily captured or appropriated by the countries in possession of tropical forests.

In the case of greenhouse gas emissions from deforestation, there are now several estimates of the minimum economic damage done by global warming, leaving aside catastrophic events. Fankhauser (1994) suggests a 'central' value of US$20 of damage for every tonne of carbon released. This implies that converting open forest to agriculture or pasture would result in global warming damage of, say, US$600–1000 per hectare; conversion of closed secondary forest would cause damage of US$2000–3000 per hectare; and conversion of primary forest to agriculture would give rise to damage of about US$4000–4400 per hectare (Brown and Pearce, 1994b). But how do these estimates compare to the development benefits of land use conversion? In the Amazon region of Brazil, R. Schneider (1992) reports upper bound values of US$300 per hectare for land in Rondônia. The figures suggest carbon credit values 2–15 times the price of land in Rondônia. These 'carbon credits' also compare favourably with the value of forest land for timber in, say, Indonesia where estimates are of the order of US$2000–2500 per hectare. The land is worth US$300 per hectare to the forest colonist but several times this to the world at large. Some of the implications of this finding are discussed in a later section of this chapter, and also in Chapter 8.

Intervention failure or government failure is the deliberate intervention by governments in the working of market forces. This can co-exist with market failure: both forces are at work at the same time. Examples of intervention failures have already been noted in this chapter, and they include the subsidies to forest conversion for livestock in Brazil up to the end of the 1980s; the failure to tax logging companies sufficiently, giving them an incentive to expand their activities even further; the encouragement of inefficient domestic wood processing industries, effectively raising the ratio of logs, and hence deforestation, to wood product, and so on. What intervention does is to distort the competitive playing field. Governments effectively subsidise the rate of return to land conversion, tilting the economic balance against conservation. For example, Ledec's (1992) study of Panama found that subsidised credit for cattle ranching is a significant factor in promoting deforestation for cattle pasture expansion. It is not the only factor and other incentives to deforest are provided by the beef and dairy markets, the securing of land claims, land price speculation, tax advantages, as well as the prestige value of cattle ranching. Deforestation is affected by other government policies including rural road construction and improvement, land titling laws and procedures, income and land taxation, beef pricing and export policies, as well as other policies which affect rural employment and poverty. Some 7–10 percent of Panama's annual deforestation was directly attributable to cattle credit programmes of the two government banks, the BNP and the BDA. These banks receive a large proportion of their lendable capital through international financing and, in particular, loans from the Inter-American Development Bank and the World Bank, so there are links with the international debt system. Further, Ledec contends that if the same analytical method were simply extrapolated to commercial bank cattle credit, this would account for an additional 13–18 percent of deforestation each year.

Table 5.4 assembles some information on the scale of the distortions that governments introduce. Such distortions are widespread. The general rule in developing countries is for agriculture to be *taxed* not subsidised, but significant subsidies exist in several major tropical forest countries such as Brazil and Mexico. By comparison, OECD countries are actually worse at subsidising agriculture. In 1992, OECD subsidies exceeded US$180 billion (OECD, 1993). These subsidies work in two ways. Subsidies in developing countries will tend to encourage extensification of agriculture into forested areas. Subsidies in the developed world make it impossible for the developing world to compete properly on international markets, locking them into primitive agricultural practices. While the removal of OECD country subsidies would appear to be a recipe for expanding land conversion in the developing world to capture the larger market, the demands of a rich overseas market are more likely to result in agricultural intensification and hence reduced pressure on forested land. Table 5.4 also shows that many developing countries fail to tax logging companies adequately, thus generating larger 'rents' for loggers. The larger rents have two

Table 5.4   Economic distortions to land conversion

Changes in agricultural price subsidies [*]		
Mexico	mid 1980s	+54%
Brazil	mid 1980s	+10%
South Korea	mid 1980s	+55%
Sub-Saharan Africa	mid 1980s	+9%
OECD	1992	+44%
Changes in timber stumpage fees as % replacement costs[†]		
Ethiopia	late 1980s	+23%
Kenya	late 1980s	+14%
Ivory Coast	late 1980s	+13%
Sudan	late 1980s	+4%
Senegal	late 1980s	+2%
Niger	late 1980s	+1%
Changes in timber charges as % of total rents		
Indonesia	early 1980s	+33%
Philippines	early 1980s	+11%

Source: Pearce and Brown (1994).
[*]Producer subsidies are measured by the 'producer subsidy equivalent' (PSE) which is defined as the value of all transfers to the agricultural sector in the form of price support to farmers, any direct payments to farmers and any reductions in agricultural input costs through subsidies (see also Chapter 6). These payments are shown here as a percentage of the total value of agricultural production valued at domestic prices.
[†]A stumpage fee is the rate charged to logging companies for standing timber. It is expressed here as a percentage of the cost of reforesting and as a percentage of total rents.

effects: they attract more loggers and they encourage existing loggers to expand their concessions and, indeed, to do both by persuading the host countries to give them concessions. Persuasion involves the whole menu of usual mechanisms, including corruption.

Macro-economic policies: debt, structural adjustment

The debt crisis has its origins in the heavy borrowing of countries to service trade deficits caused by the first oil price shock in the early 1970s. It was enhanced in the 1980s by the decision of international agencies to finance indebted countries in the hope that encouraging economic growth would spring these countries into long term growth and the debt would be repaid eventually. However, worsening terms of trade for export commodities and defaulting on debt repayments led to a breakdown in the system. It is often postulated that the huge level of external debt in many developing countries since the mid-1970s onwards has contributed to creating the environment where deforestation occurs. The mechanisms by which this occurs are:

(1) in creating high domestic demand for foreign exchange to pay back the debt, which is satiated through the export of timber and other internationally tradeable products;
(2) in creating a macro-economic environment generally deleterious to economic growth and hence forcing people into the extensive use of marginal lands; and
(3) by forcing governments into a position where they reduce expenditure, especially on environmental protection and other services.

However, these mechanisms are by no means straightforward and any simple correlation between indebtedness and deforestation rates is spurious, due to the effects of *scale*, as proved by Gullison and Losos (1993). For example, although countries such as Brazil have high levels of both debt and deforestation (in terms of hectares lost), when these variables are standardised and taken on a per capita basis, there is no correlation across countries between debt and deforestation.

The countries of Latin America did not significantly raise their exports of timber, or of beef (the hamburger connection) in the 1980s in response to rising debt, according to Gullison and Losos (1993). The demand for these products may be an important contributory factor to deforestation in countries such as Brazil, but this is not specifically exacerbated by high levels of debt and is more closely related to domestic policies. The impact of indebtedness on government expenditure may directly lead to reductions in expenditure on environmental protection in its narrowest sense, but more importantly make the poorest groups in society worse off. Expenditure on health, agriculture, education and public services all fell during the 1980s across Africa while interest payments on debt rose (Stewart, 1992).

It is in these circumstances that exploitation of open access resources and marginal land occurs. Additionally the lower the income of a country, the less it is likely to spend on direct environmental protection—an effect known as the environmental Kuznets effect. Confronted with falling living standards, an indebted nation may find it preferable to release resources previously devoted to environmental protection for the purpose of boosting production. The fact that during periods of fiscal reductions, the budget for Mexico's Bureau of Urban Development and Environment fell faster than government spending in general is consistent with the operation of a Kuznets curve effect (Pearce et al., 1994).

Debt may therefore have a primarily indirect effect on rates of deforestation as suggested by Kahn and MacDonald (1992) in a behavioral model which illustrates the effects of macro-economic conditions and political stability. This suggests that debt may lead to myopic behaviour where deforestation accelerates beyond an optimal level to generate income to meet short term needs at the expense of future consumption. Deacon and Murphy (1992), on the other hand, argue that debt does not cause deforestation, but rather that debt and deforestation are symptoms of the same myopia; they suggest political instability as a potential source of this myopia. In a later study, Kahn and MacDonald (1994) present empirical results which suggest that there is a strong statistical relationship between debt and deforestation; on the other hand, Shafik's (1994) study indicates that there are very few macro-economic causes of deforestation at the aggregate level.

Since 1989 there has been a rethinking of international financing of debt, so that debt forgiveness is presently being considered rather than continued debt financing (Krugman, 1992). Debt relief costing only a fraction of what was suggested at UNCED in 1992 to be necessary for sustainable development worldwide, would eliminate this contributing factor to poverty, environmental degradation and deforestation, in particular in the developing indebted countries (Brown et al., 1993).

Many developing countries even if not indebted face chronic problems of stagnant economies and underemployment. In the 1980s multilateral aid agencies, primarily the World Bank and the International Monetary Fund, financed reform of economies in developing countries on the condition that structural reforms were undertaken (Cornia et al., 1992 on the African experience for example).

Environmental pressure groups have highlighted the role of these Structural Adjustment Programmes (SAP) in accelerating deforestation; however, the empirical evidence is mixed, and the effects of policies will depend on the exact package of measures adopted. For example, policies which remove subsidies from farm inputs (pesticides and fertilisers), as adopted by Mexico, or which tax inputs (e.g. Thailand) may encourage extensification rather than intensification of land use (Reed, 1992). The case studies in the Reed volume illustrate the lack of a clear methodology for analysing the effects of SAPs on the environment in general (Cromwell and Winpenny, 1993), and also how policies aimed at improving economic performance in one sector of the economy (e.g. smallholder production of export crops) may have unwelcome side-effects in another (deforestation). While conventional wisdom and proponents of SAPs maintain that stabilisation and adjustment programmes can also benefit environmental management in so far as they improve macro-economic stability, lengthen planning horizons and improve the workings of the price mechanism, experience to date is more varied and in many cases the effects on forest management are negative (see for example Knerr (1992) on the effects of SAPs on deforestation in sub-Saharan Africa). Much criticism of SAPs has also focused on their effects on income distribution and on poorer sections of the population in developing countries, and this is also likely to affect deforestation, through links with poverty and landlessness.

Summary: The causes of tropical deforestation

The discussion in the last two sections has highlighted the range of different factors affecting the rate of deforestation. The direct cause of deforestation is the conversion of land to other uses, predominantly agriculture, but the underlying forces driving this conversion are far more complex. Simplistic explanations, blaming deforestation on high population growth rates in developing countries, the demand for tropical timbers in the North, or pressures to meet debt repayments, are found to be far from satisfactory. Various studies in recent years have failed to provide conclusive answers applicable to a range of different circumstances. They have also served to highlight the synergistic effects and interconnectedness of the different factors, e.g. logging opening up areas for settlement and clearance for agriculture.

Table 5.5 provides a brief summary of some of these studies, indicating the variables and methods examined. What becomes apparent on reviewing the evidence, is that a number of different economic and policy interventions enacted in the tropical countries themselves, through the workings of the international system, or even by importer countries, will affect the way in which forest resources are managed. At the present time, many of these policies provide incentives to deforest—either to convert forest land to other uses, or to harvest timber at an unsustainable rate and by methods that cause degradation and eventual destruction—and very few provide incentives to conserve forests. It is this finding which lies at the heart of implementing measures to slow deforestation and encourage sustainable management practices.

5.6   SLOWING THE RATE OF TROPICAL DEFORESTATION

Having discussed the different causes of tropical deforestation, we now need to consider whether it is better to tackle the direct or underlying causes, or both at the same time. In doing this we must be aware of the trade-offs, particularly for poorer developing countries, and the competing demands for economic development in diverting scarce resources to encourage forest conservation. Do we address the symptoms or causes of deforestation? It is clearly not enough to designate protected areas, a policy which has so often failed or been ineffective, or even had negative results in terms of well-being of local people. In addition, many developing countries do not have sufficient resources to effectively enforce protected area boundaries. Conservation strategies need to address causes therefore, not symptoms. But which causes? If the premise is taken that deforestation is an inevitable consequence of social and economic development then deforestation needs to be controlled by increasing farm productivity and sustainability, and by concentrating the most intensive farming practices in the most fertile land. This implies encouraging Boserupian intensification in the belief that this will displace the need for extensification and the conversion of forest. Specifically, efforts in rural development, agricultural intensification and creating employment opportunities in rural areas are required in order to relieve the need for expansion into forests. An alternative set of policy measures is based

Table 5.5 Selected econometric studies on deforestation

Study	Type of analysis	Dependent variable	Independent variables					Other significant variables
			Population	Population density	Income	Agricultural productivity	External indebtedness	
Shafik (1994)	Panel regression of the causes of deforestation, three models	Annual rate of deforestation 1962–1986, 66 countries			GDP per capita not sig	GDP per capita not sig	Debt per capita, not sig	Investment rate—positive Electricity tariff—negative Trade shares in GDP—negative Political rights—positive Civil rights—positive
		Total deforestation, 1961–1986, 77 countries			not sig		not sig	
Burgess (1992)	Cross-sectional analysis of def. in 53 countries	Change in closed forest area, 1980–1985		negative 0.01	Real GNP per capita in 1980 positive 0.05			Roundwood production per capital 1980—negative 0.05 Log of closed forest area as a percentage of total forest area in 1980—positive 0.1
Burgess (1991)	Cross-sectional analysis of def. in 44 countries	Model 1: level of deforestation	Population growth negative 0.05		GDP per capita positive 0.05		debt-service ratio as a % of exports positive 0.05	total roundwood production—positive 0.05
		Model 2: level of deforestation	negative 0.01				positive 0.10	food production per capita—positive 0.10 total roundwood production—positive 0.05

Study	Method	Dependent variable	Population	Per capita income	Sign	Other variables
Southgate (1994)	OLS regression analysis of causes of agricultural colonisation in 23 Latin American countries.	Growth in the area used to produce crops and livestock	Population growth—positive 0.01		negative 0.01	Agricultural export growth—positive 0.05
Kahn and MacDonald (1994)	2-Stage-least-squares model to show economic mechanisms by which debt may lead to deforestation	def. area (1000 ha)	negative 0.017		positive 0.01	Forested land area—positive 0.01 Annual change in public external debt—positive 0.05
Capistrano and Kiker (1990)	OLS analysis of macro-economic factors of deforestation in 45 countries 1967–1985, two linear models	depletion of broadleaved forests during four periods, 1967–1971 (P1) 1972–1975 (P2) 1976–1980 (P3) 1981–1985 (P4)	positive 0.01 P2	GNP per capita positive 0.01 P2, P3	Debt-service ratio negative 0.01 P2	Log export value—positive 0.01 P1 Real devaluation rate—positive 0.01 P3 Cereal self-sufficiency ratio—positive 0.01 P2 Arable land per agricultural capita—positive 0.01 P4
Rudel (1989)	Decline in closed tropical forests for 36 countries across Africa, Asia and Latin	average annual decline in hectares of a country's tropical forests during the period 1976–80	Population growth— positive 0.001 Rural population growth— positive 0.01	GDP per capita positive 0.01		Forest land area—positive 0.001

Table 5.5 Table 5.5 (continued)

Study	Type of analysis	Dependent variable	Independent variables					Other significant variables
			Population	Population density	Income	Agricultural productivity	External indebtedness	
Palo et al. (1987)	Cross-sectional test of factors influencing deforestation in 72 countries	absolute forest cover in 1980.		negative 0.01				Food production per capita—negative 0.01 Share of forest fallow—positive 0.1 Agricultural area coverage—negative 0.1
Allen and Barnes (1985)	Deforestation between 1968 and 1978 in 39 countries in Africa, Latin America and Asia	Model 1. annual change in forest areas Model 2: the decade change in forest area, 1968–1978	pop. growth negative 0.10					Logarithm of %of forest cover 1986—positive 0.10. The % area under plantation crops in 1968—negative 0.05 Per capita wood fuels consumption and wood exports in 1968—negative 0.05
Lugo et al. (1981)	Deforestation in all greater Caribbean countries	%forest cover		negative 0.001				Energy use per unit area—positive 0.001
Reis and Guzman (1994)	Brazilian Amazon deforestation and its contribution to CO_2 emissions	Deforestation density				positive 0.001		Cattle herd—positive 0.05 Logging—negative 0.01

Reference	Study	Dependent variable				Independent variables
Katila (1992)	Deforestation in Thailand	Relative forest cover	negative 0.01		negative 0.05	Wholesale price of construction timber—negative 0.01.
Constantino and Ingram (1990)	Deforestation rates in Indonesia	Relative forest cover	negative 0.01		positive 0.01 (rice production used as a proxy)	Time—negative 0.01
Kummer and Sham (1994)	Deforestation in post-war Philippines	Cross-sectional analysis for the years 1957, 1970 and 1980. Absolute amount of forest cover per province in hectares.	negative 0.05 1970 and 1980	GDP per capita positive 0.05		Road density—negative 0.05, 1957, 1970 and 1980 Kilometres of road—positive 0.05 1980
		Panel analysis of absolute loss of forest cover, 1970–1980 per province in hectares				Forest area—positive 0.05 Distance from Manila—positive 0.05 Logging in 1970—positive 0.05.
Panayotou and Sungsuwan (1994)	Deforestation in north-east Thailand	Forest cover	negative 0.01	positive 0.01 (provincial income)		Wood prices—negative 0.01 Distance from Bangkok—positive 0.01 Rural roads—negative 0.10 Rice yields—positive 0.10 Price of kerosene—negative 0.01
Southgate et al. (1989)	Deforestation in 20 cantons in eastern Ecuador in early 1980s	Deforestation	(agricultural population) positive 0.05			Tenure security—negative 0.06

Source: Brown and Pearce (1994a).
Notes: Negative and positive means the independent variable is negatively (or positively) correlated with the dependent variable at the significance level given.

on the notion of market failure as a cause of deforestation: that the value of forests is unaccounted for in private decisions. Policies therefore involve optimising the value of forests and making forest conservation pay on a local, national and international level. These include multiple use management (Panayotou and Ashton, 1993), setting up extractive reserves, and enhancing tourism potential. Each of the different options are briefly considered below.

Agriculture and rural development

Since agricultural expansion is a major cause of tropical deforestation, improving agricultural productivity should be a core part of any strategy to limit deforestation. According to Grainger (1993), such a programme should focus on three aspects of agriculture: first, improving overall land use policy and planning; secondly, improving shifting agriculture practices; and thirdly, introducing measures aimed at increasing the productivity of permanent agriculture. These are essentially technical solutions to the problem of deforestation based on: scientific improvements in agriculture and planning; more intensive systems of short-rotation shifting cultivation which maximise soil fertility and utilise agro-forestry and tree gardens, for example; or increasing production from permanent agriculture in more fertile areas. These strategies, however, are dependent on economic and social determinants of land use. If rural development is to take place, the institutions on which sustainable resource use are founded need to be strengthened. The breakdown of institutions (of land tenure, division of labour, and of predictable stable markets) often triggers action leading to environmental degradation (Kates and Haarmann, 1992; Lopez, 1992). Social and institutional stengthening, in its widest sense, is therefore more important than technical strategies for rural development.

Perhaps the most important of these policy areas concerns property rights and land reform. Land tenure is a critical determinant of peasant encroachment into forests; in many tropical countries inequitable distribution of existing prime agricultural lands induces peasants to migrate to marginal forest lands and practise slash-and-burn cultivation. For example, in Guatemala, a country experiencing rapid deforestation, 2.2 percent of the population owns 70 percent of the land suitable for permanent agriculture. Eighty-five percent of Guatemala's rural families either own no land whatsoever, or own too little to support their subsistence needs (Ledec, 1985). A recent volume edited by Colchester and Lohmann (1993) argues that agrarian reform is the only strategy that will control deforestation in the long term, and that it is the inequitable distribution of resources and power in many developing countries that causes not only deforestation, but also widespread and abject poverty and high rates of population growth: 'to attribute tropical deforestation to population pressure is to argue that spots cause measles since the two are joint manifestations of exploitative social relations' (Westoby, 1989).

Again, the implementation of such policies may engender trade-offs perhaps between indigenous forest-dwellers and agricultural settlers. Historically, indigenous people have often been discriminated against and persecuted, and government policy has supported the interests of other groups in terms of land reform. Perhaps things are slowly beginning to change; indeed 1993 was deemed the Year of Indigenous People by the United Nations. Multiple use management strategies, as explained below, may be able to support the livelihoods of indigenous people especially, although only if reinforced by validation of indigenous people's property rights. There is now a growing school of thought which argues that the best way to conserve areas of tropical forest is to base policies on the institutions and utilisation of forests by traditional peoples themselves (see Redford and Padoch, (1992), although Richards (1992) demonstrates that this may not be appropriate in all cultural situations). We will return to this theme in following sections.

Forest management

Improvements in forest management, which make forests more valuable assets and therefore provide incentives for their conservation, can be made in a number of different ways. In the first place, the implementation of selective logging practices can be encouraged. This involves strengthening the management and increasing the profitability of existing logged areas which, in theory at least, and given the appropriate policy climate, should remove incentives to expand into new areas, and all the accompanying dangers of encroachment that go along with forest development. Selective logging operations would be improved by the introduction of better monitoring and regulation, longer concession periods, strengthened protection of forest areas, and improved logging techniques (Vanclay, 1992; Grainger, 1993).

The concept of multiple use of forests is based on the recognition that a variety of goods and services can be produced from the same land, either simultaneously or serially, and that this management can greatly increase the net value of the forest. This was shown in the discussion of total economic value earlier in this chapter. In fact this approach can help ensure conditions for the sustainable production of timber. Panayotou and Ashton (1993) distinguish six basic uses of forest lands: timber production; the production of non-timber forest products; provision of environmental services such as soil and watershed protection, and conservation of genetic diversity; regulation of climate and carbon sequestration; recreational and aesthetic benefits, including tourism; and conversion to other uses, for instance agriculture and livestock production. Multiple use management aims to optimise the mix of these uses.

The successful implementation of multiple use management and the extent to which its potential can be realised will depend on institutional reform, government policy, international co-operation, and further research to fill gaps in knowledge.

Perhaps the most important of these is the clarification of property rights, and general uncertainty and insecurity of ownership of forest resources is the greatest restraint to optimal forest investment according to Panayotou and Ashton. Historically, most tropical forests were communal or tribal property to which members of the community or tribe have had customary rights of access and use. However, in the last century over 80 percent of the forests of tropical Africa, Asia and Latin America have been brought under government ownership, with special legal status given to particular areas such as forest reserves and national parks. Governments have not generally been able to enforce their ownership, and deforestation has been the result. The solution lies in establishing well-defined property rights and these may be based on the local institutions and customary rights of indigenous people and forest-dwellers, and may not constitute outright ownership, but more complex patterns of access and use of different resources (Fortmann and Bruce, 1988). These strategies may also be able to provide sources of livelihood for forest-dwellers. Panayotou and Ashton (1993) suggest the following guidelines for establishing property rights to enhance possible success of multiple use management.

— Forestlands with no significant externalities can be safely distributed and securely titled to individuals.
— Forestlands with localised externalities, such as local watersheds and extractive reserves, can be made communal property.
— Forestlands with national or international externalities, such as major watersheds and nature reserves, should remain under state ownership.

There is one other way of utilising multiple products from tropical forests whilst at the same time ensuring that their biological diversity and other ecological functions are not disrupted; so-called 'extractive reserves'. Extractive reserves have been defined as 'forest areas inhabited by extractive populations granted longterm usufruct rights to forest resources which they collectively manage' (Schwartzman, 1989, p. 151) and have been pioneered in the Amazon but replicated in many other parts of the world including Indonesia (Peluso, 1992), Samoa (Cox and Elmqvist, 1991), and Guatemala (Nations, 1992). These initiatives may be supported by efforts to find new markets for rainforest products and thus generate income from sustainable extractive activities (Plotkin and Famolare, 1992).

Reversing policy failures

Having identified the government and policy failures which provide incentives to deforest, it might be easy to assume that these distortions would be simply rectified by reversing policies; for example, removing subsidies to cattle ranching and adjusting taxes. There is undoubtedly much that can be done by national governments in this respect to stem deforestation; however, there are also a number of factors which militate against the implementation of such straight-forward measures. These include the pressures from powerful interest groups, such as large landowners and industry who profit from deforestation, and international pressure from importer countries. In addition, many governments are dogged by short term expedients for re-election, and so may favour profitable exploitative strategies over long term sustainability. The political and economic marginalisation of indigenous forest-dwellers means that very often, unless supported by international pressure, their interests are not fairly represented in national politics. Many different factors therefore mean that it is not as simple to reverse destructive policies as it may appear. Following the discussions in this chapter on how economic, political and institutional factors are the key to understanding and hence slowing deforestation, Table 5.6 presents a list of key policy changes which follow from these. The policy changes represent broad policy directions which are generalised, rather than specific recommendations which will of course be tailored to the particular circumstances of individual countries.

The international dimension

There is of course an international dimension to the policy measures required to stem tropical deforestation and encourage sustainable forest management practices, as illustrated by Table 5.6. More specifically, a number of initiatives have been implemented at a transnational level with the intention of conserving

Table 5.6   Key policy changes for controlling deforestation

Priorities for governments of developing countries
  Increased support for small farmers through rationalisation of agricultural policy
  Strengthening of forestry departments
  Revision of concession agreements and fees
  Development of a national conservation strategy
  Improvement of monitoring of natural resources
  Increased environmental component of national development strategies
  Reformation of land tenure recognising communal management systems

Priorities for governments of developed countries
  Debt relief and funding of positive development strategies
  Support new conservation funding mechanisms

Priorities for international agencies
  Development of national accounting which reflects resource depletion
  Establishment of a continuous global satellite monitoring system for the tropical forests
  Implementation of mechanisms for local appropriation of global benefits in context of Climate Change and Biodiversity Conventions

---

tropical forests. These include international agreements on forestry; reforms of the international trade in tropical timber including timber certification schemes; green conditionality and Debt-for-Nature Swaps. In addition, big international non-governmental organisations (BINGOs), such as Greenpeace International and Worldwide Fund for Nature, put pressure on both exporters and importers of tropical timber to ensure timber comes from sustainable sources. These NGOs have done much to raise public awareness of some aspects of tropical forests and their loss.

5.7 CONCLUSIONS

High rates of tropical deforestation are essentially a result of human action. Although natural phenomena may cause loss of forests, almost all of the deforestation experienced globally is caused by humans, constituting the clearing of forest to put land to other uses. In the future, we may see dramatic changes in forest vegetation due to the effects of climate change or global warming; again, this will be human-induced and as a result of our inability to act decisively now to combat the so-called greenhouse effect.

A fundamental feature of excess deforestation is that its causal factors are linked together as various chains or mechanisms into a causal system (Palo, 1990). The mechanisms comprise mostly positive feedback loops which tend to accelerate the process of deforestation. This therefore makes the design of policy to slow deforestation very complex and there is a need to understand these causal relationships before effective policy can be implemented. Our analysis has shown that the introduction of new land use may be the motive for deforestation, but that to understand why it occurs, we need to examine underlying causes. These inevitably rest with national governments and policy, and through the workings of the international economic and political system.

This chapter has taken as an implicit assumption that the conservation of existing tropical forests is a primary mechanism to reduce greenhouse gas emissions from the land use sector. From a carbon sequestration perspective, depending on how existing biomass is disposed of (i.e. whether it is burned or otherwise), it may be desirable in certain circumstances to replace these forests, most of which consist of relatively unproductive so-called climax vegetation, with plantations of fast-growing monocultures. Such a strategy would ensure the highest rate of active carbon sequestration (as discussed in Chapter 3). However, what this chapter has stressed is that many other benefits of tropical forests risk being lost if the current high rates of destruction continue for any length of time. These benefits include being the home and means of livelihood for forest-dwellers and other people; being the habitat for great numbers of animal, plant and micro-organism species, many of which are still unknown to science; and providing a range of micro- and macro-level ecological services, including soil conservation and fertility, water cycling and climate control. New plantations may be able to provide some of these benefits, but not all of them, not to such an extent as existing forests. Sustainable forest management policy should aim to maximise these multiple benefits so they can be enjoyed by present and future generations.

CHAPTER 6

Agricultural Policy to Reduce Methane Emissions

6.1 INTRODUCTION

Are there opportunities to reduce greenhouse gas emissions in the agricultural sector? Any opportunities seem to be based either on changing the *pattern of land use*, or on changing the *management* of land in its present use. Thus in the previous chapter we discussed how the conversion of land to forestry can be stemmed, thereby reducing emissions. This chapter investigates how management practices and the use of agricultural land influence greenhouse gas emissions and if there are opportunities to reduce these. Specifically, methane emissions from livestock and from rice growing are examined. Agricultural policies in the countries of the European Union ensure high levels of support for agricultural and livestock production. It is shown that the reduction of CH_4 emissions is a significant but previously unaccounted for benefit of the reform of these agricultural policies. In rice production, by contrast, opportunities for reducing production are severely limited by present and future demands for rice as a staple food. Options for the reduction of CH_4 emissions through changing technology and management practices are available in both the livestock and the rice cases, but their adoption depends on farm-level decisions, based on the prevailing technical and policy environment.

6.2 METHANE EMISSIONS FROM LIVESTOCK

As is shown in Part II of the book, CH_4 is produced through biological decomposition by micro-organisms under anaerobic conditions, such as exist in swamps and paddy fields, in peaty soils, in the guts of livestock, in sewage and landfill sites and as produced by termites. Industrial sources include coal mining, the flaring of natural gas in oil production, industrial processes and the inefficient combustion of hydrocarbon fuels. It is important that CH_4 be considered as part of a comprehensive strategy of greenhouse gas abatement, not only because of its higher radiative forcing, but also because of the prospect of the greater future impact, both due to the reduction over time of the capability of the atmospheric sink and of the soil sink to absorb increased concentrations.

Emissions of CH_4 from ruminating animals and the slurry associated with their intensive management are relatively well quantified, and the options for reducing emissions (either by management regime changes or by reducing herd size) are clear. The atmospheric concentration of CH_4 is presently more than double its pre-industrial level, now estimated at 1.72 ppm by volume. The rate of increase has slowed in the last decade, either due to changes in the emissions from anthropogenic sources or to the increased concentration of OH radicals in the troposphere.

Across all the sources, agriculture-related emissions form almost half the anthropogenic emissions (even excluding the land use change related emissions from biomass burning). Emissions from ruminants are estimated to provide an atmospheric source of 65–100 mt CH_4 year^{-1}, based on the estimates of emissions per animal of Crutzen et al. (1986) and summed for the global ruminant animal population by Lerner et al. (1988). Results for selected countries are shown in Table 6.1.

The results in Table 6.1 show that many countries with low per capita incomes

Table 6.1   Aggregate methane emissions from domestic animals for selected countries for 1984

Country	Emissions mt CH_4 year^{-1}	% of total
India	10.27	13.5
USSR	8.05	10.6
Brazil	7.46	9.8
United States	6.99	9.2
China	4.37	5.8
Argentina	3.11	4.1
Australia	1.90	2.5
Pakistan	1.54	2.0
France	1.52	2.0
Mexico	1.51	2.0
Bangladesh	1.43	1.9
Ethiopia	1.20	1.6
United Kingdom[*]	1.17	1.5
Sudan	1.00	1.3
Others	24.31	32.1
Total	75.83	100.0

Source: Lerner et al. (1988), based on FAO country animal data, and own calculations for UK based on MAFF data.
[*]The UK estimate is included for comparison but does not employ the same methodology as Lerner et al.; The UK estimate is estimated later in this chapter (see text).

are among the largest emitters due to their large ruminant populations. Livestock herds often constitute a principal source of income and wealth in some countries: reduction of emissions through reducing herd size would therefore be infeasible; neither would it be equitable (see Drennen and Chapman (1992) on this point). The handling of animal wastes globally adds almost one-third to the output of domestic animal emissions (Watson et al., 1992). This is largely in industrialised countries where intensive management leads to an increased source capacity with the same herd size. Intuitively, therefore, this is an area where emission reduction is feasible.

The next part of the chapter discusses the *value to society* of reducing CH_4 emissions, and the private cost parameters involved in the agricultural sector. The social value of emission reduction is in terms of the value of global warming avoided. This is derived from economic studies of global warming impacts which have almost exclusively defined the cost of CO_2 emissions in monetary terms per tonne of carbon. To translate these estimates to CH_4 equivalent requires comparison of the relative global warming impacts of the different trace gases. In physical radiative forcing terms, the set of global warming potentials (GWPs) described in Chapter 1 relates the gas to CO_2.

The analysis in this chapter is restricted to the UK agricultural sector context. The global value of reducing emissions is estimated, and estimates of cost-effectiveness are presented. The Common Agricultural Policy of the European Union has operated a system of price support in the past 40 years in Europe that has encouraged expansion of agricultural output to gain food self-sufficiency, but levels of output in many commodities are now greater than domestic European consumption requirements. This situation is repeated in many industrialised countries and regions such as the US and Japan, and the consequences of this on external economies, on the rural environments of the domestic countries, and on the budgetary costs of supporting agriculture are great. It is predominantly these costs and the impetus for the freeing of agricultural markets within the GATT negotiations that have led to recent policy reform in agriculture.

The benefits of these reforms on the cycles of the major greenhouse gases has not generally been addressed, at least in the European Union. Land use change stimulated by agricultural development can be shown to contribute to net emissions of greenhouse gases in a temperate country such as the UK (Adger et al., 1992b). In its latest reforms, the EU is considering subsidising afforestation to enhance carbon sinks. These policies are discussed in Chapter 7.

Agricultural emissions of greenhouse gases are also affected by agricultural practices; for example, the more intensive the management regime of livestock in terms of stored feed and nutrient supplements and in terms of stored slurry, the greater the related emissions.

More intensive production systems result in emissions of greenhouse gases from a number of different processes. Principally, the anaerobic decomposition of manure produces CH_4 in addition to that produced from enteric fermentation.

This is likely to add 5–10 percent to methane emissions as calculated in this study and confirmed for the Netherlands by Swart et al. (1993). Secondly, intensive systems are likely to use higher levels of inorganic fertilisers of grasslands, which will result in N_2O emissions. In addition, intensive systems cause other pollution in the form of run-off from fertiliser application, slurry and silage sewage.

Price support in the EU has raised the intensity and scale of agricultural production. This is shown by various indicators for the UK for example, such as the input of fertilisers which increased by over 30 percent between 1978 and 1988, and the areas of tillage, which increased by 11 percent in the 20 years post-1973. This has resulted in loss of grassland and semi-natural vegetation (Adger, 1993). The cost to both taxpayers in terms of expenditure on price support, and consumers in terms of higher food prices and perceived environmental degradation of rural resources and landscape, has led to pressure for reform of the system of support in western Europe. The picture is repeated elsewhere in the world, with the other main agriculture-supporting countries, Japan and the US, being urged to reduce domestic agricultural support to producers to promote greater access to markets for producers in developing countries and the Pacific rim. However, domestically in the US, Japan and internally in the EC, agriculture has received state support for centuries and a vociferous and powerful political lobby ensures that the mechanisms of support change only slowly. Despite this, in the EU it is the former issues, of cost to taxpayer and consumer, which have been the driving force for reform. If the price support system were to move to a different basis of support, and took environmental protection as a central objective, then much of the reform could promote goals such as the reduction of pollution and emission of greenhouse gases.

6.3 THE ECONOMICS OF METHANE EMISSIONS

Global warming abatement confers value to society by avoiding the impacts of global warming, some of which may be beneficial but some of which may be catastrophic, but which in general are assumed to be negative and not known with certainty. This value is generally unaccounted for in decisions concerning resource use such as in the agricultural sector. However, as signatory parties to the Climate Change Convention are now committed to reducing their emissions of all greenhouse gases, the importance of these emissions should be recognised and least-cost strategies to reduce emissions need to be identified. The value to society of reducing emissions from the UK livestock sector is now illustrated.

A framework for analysis

In a cost–benefit framework, the net worth (the difference between benefits and costs) of decreasing livestock price support is the value of the benefits of a reduction in pollutants less the costs, i.e. the reduction in income less the budgetary and consumer costs of such a reduction. This calculation has to be carried out over present and future time periods, because although the costs of price support loss will be felt immediately by farmers, some of the benefits of reducing emissions will not occur till later time periods. The conventional economic method for expressing costs and benefits accrued in different time periods is 'net present value', which reflects the generally held view that the same monetary amount (say £1) is worth more now, than the prospect of £1 some time in the future. The net present value then discounts the future value at a rate which can be empirically observed by economic behaviour in an economy. This net value expressed in total value over time can be written algebraically for the reduction of price support to agricultural producers:

$$\text{Net present value} = \sum_0^T \frac{(B_t - C_t)}{(1 - r)^t}$$

where:

B_t = value of benefits of pollution avoided (£),

C_t = net value of costs of lost income less the saved government budgetary costs (£),

T = time horizon (∞),

t = time period in which costs or benefits accrue,

r = rate of discount.

The benefits (B_t) here are those associated with methane emissions from livestock. Ideally B_t would include acid rain damage through ammonia volatilisation (NH_3) and nitrate (NO_3) leaching costs on water quality at the time which they occur. The benefit estimates are then a lower bound estimate of the environmental benefits of the reduction of the aggregate herd size. C_t is the projected social value of production lost as a result of reduced market support. The net benefit is treated as an annuity to be discounted in perpetuity (i.e. $T = \infty$).

The issue of a discount rate applicable in the context of assessing both national and global benefits remains to be resolved. However, it seems appropriate that the relevant rate would approximate some global rate of social time preference giving adequate weight to future damage and risks of unforseen catastrophes. The sensitivity of abatement scenarios to various rates of time preference has been explored in Cline (1992). A discount rate of three percent is used in the analysis here, though the estimates of damage due to global warming from the studies cited possibly justify lower discount rates on various grounds (discussed below).

Costs

The costs of agricultural policy reform in this benefit–cost analysis are the income foregone in perpetuity, as the change considered is a permanent change in the overall stock. However, the *financial* losses to the farmer are greater. The short run estimate of the cost of reduction of herd size is the revenue lost less the variable costs not incurred (i.e. the margin between revenue and saved variable and operating costs). Financial analysis of lost income due to policy reform does not reflect the *social opportunity costs* of reduction in agricultural income (C_t in the equation above), due to the government intervention in agricultural markets. Price support in agriculture has many objectives. Among the objectives in the EU are the maintenance of agricultural landscapes and of the structure of rural society. The present price support also has numerous negative externalities on countries outside the EU and on the rural environment within it, including increased agricultural pollution (Hanley, 1991) and wetland loss (Turner and Jones, 1991); increased instability and decreased prices in world food commodity markets; and reduced access to protected markets for developing country agricultural exporters (Buckwell, 1991).

In this study we do not account for these various externalities, both positive and negative, but in line with the principles of social cost-benefit analysis, we adjust distorted agricultural support for government transfers. The simplest mechanism to try to correct for these is to use world market prices as the true value of agricultural output. However, world market prices themselves are distorted by export subsidies from major agricultural producing regions and other trade barriers.

The EU's Common Agricultural Policy (CAP) has a complicated system of price support for all agricultural products. This is based on a set intervention price (higher than world market prices) at which surplus produce is centrally purchased and up to which level exports are subsidised. The effect of this support on margins (revenue less costs) has been mixed as higher land, capital and input prices have resulted from the guaranteed intervention pricing in the last 40 years of the CAP. Escalating budgetary costs of the CAP, and the inclusion of agriculture in the Uruguay Round of GATT have led to pressure for reform, especially in the dairy, beef and cereals sectors.

Reforms have been given further impetus in 1992 by the so-called MacSharry Proposals. Under these proposals in the beef sector, for example, the requirement for agricultural support to be less distorting to trade has also led the reforms to impose other constraints as well as reduction of the intervention price. The intervention price is set to fall by 15 percent and additional stocking density reductions are to be phased in over four years. Incentives for further reductions in stocking densities are provided by an extensification premium. The intervention price will fall from £381 per 100 kg beef in 1992 to £324 per 100 kg by 1995 (Meat and Livestock Commission, 1992).

The real value of agricultural production when the price support is stripped away ('social value' in economic terminology) is therefore estimated by using the concept of Producer Subsidy Equivalent (PSE), which, when reported as a percentage, is the proportion of the domestic market price that is in the form of output subsidies:

$$PSE = \frac{Price\ support\ (\text{£}m)}{Value\ of\ production + Direct\ payments\ (\text{£}m)} \times 100$$

(1-PSE) can be taken as an estimate of the social value of agricultural output, and PSEs are conveniently estimated for all major agricultural products in the EC and elsewhere.

PSEs do not incorporate the trade distortionary impact of agricultural support. Some mechanisms for support such as research and development expenditure are not trade distorting. Also, in the future it is likely that agricultural support in the EU and elsewhere will not be tied directly to output, but rather to environmental 'outputs', and hence the PSE measure may become less relevant (Josling and Tangerman, 1989). A PSE measure also ignores protection of the markets for inputs, so that the world market price of agricultural products may either overestimate or underestimate the social value of agricultural output (Willis et al., 1988). Nevertheless, the PSE is a readily available and convenient measure of the level of protection of agricultural output.

The present values (discounted at 3 percent) of the social value of the revenue foregone (C_t) per head for different types of livestock are shown in Table 6.2. The revenues are based on average data (MAFF, 1993a), multiplied by the PSEs for the sectors. PSEs fluctuate year on year with support price and 'green' exchange rate changes (for the EU countries) so the most recently available five-year averages for the UK are used. As the objective of CAP reform is to reduce trade-distorting producer support, it is projected that the PSEs will fall over the coming years. The extent of support is highest in the sheepmeat sector (74 percent) and lowest in the pigmeat sector (5.8 percent). The assumption has been made here, that the PSEs in the beef, dairy and sheepmeat sectors will fall by five percent per year over the next three years and stabilise thereafter. This is a highly uncertain parameter, but it is reasonable to assume that the major part of support will remain, as this is a central objective of the present reforms.

Benefits—reduction of methane emissions

To estimate in monetary terms the benefits of reducing CH_4 emissions, three steps are necessary. First an estimate of CH_4 emission per head is required for both direct emissions and from stored slurry. To estimate the value of this, CH_4 has to be compared to other greenhouse gases using Global Warming Potentials or some other comparison. GWPs are used here whilst recognising their inherent problems. A range of monetary values of the future damage from CO_2 equivalent emissions are then given and the information is combined to produce a value per head of CH_4 emission reduction.

Table 6.2   Costs of reduced price support to UK livestock sector

	Revenue (£ per head per year)	1- Producer subsidy equivalent	Assumption of changes in PSE
Beef	282.6	1-0.552	5% fall per year for 3 years
Dairy	827.7	1-0.656	5% fall per year for 3 years
Sheep	37.3	1-0.740	5% fall per year for 3 years
Pigs	1822.6	1-0.058	No change

Sources: Adger and Moran (1993), MLC (1992), MAFF (1993a), PSE data supplied by MAFF.
Note: Present value of social opportunity cost of reduced output is calculated as:
Revenue × (1-PSE) per year discounted at 3% , and assuming changes in PSE as shown (see text).

Methane emissions

Methane is produced directly by enteric fermentation in the guts of ruminants and also from methanogenesis of stored manure. Uncertainty surrounding these emissions is less than that from land, as direct emissions are relatively easily determined experimentally. To estimate emissions on a per head basis, direct and indirect emissions must be accounted for. The procedure set out below includes an estimate of emissions from manure production from the proportion of livestock kept indoors and the proportion of manure which falls under the conditions for methane production. These rates are obviously less for sheep and pigs than for cattle, but also vary between the beef and dairy herd and with age distribution and management. Dairy cows in milk, for example, produce more methane than those not in milk (95 kg and 65 kg CH_4 per head per year respectively).

The direct emission estimates are based on reviews of average emissions across a range of feeding regimes and climatic conditions for temperate countries (Crutzen et al., 1986; Costigan, 1993). The estimates adopted are shown in column 3 of Table 6.3, and weighted, in the case of beef and dairy cattle, by the proportions of the UK herds less than two years old (for beef) or in milk (in the case of dairy).

Indirect emissions are not relevant for sheep, but are for the populations of beef and dairy cattle, and pigs. Methane is emitted from slurry mainly under conditions of less than 10 °C average temperature, which occurs only part of the year in the UK. Under optimising conditions, cattle manure can produce up to 180 kg CH_4 per head per year (Hayes et al., 1980). These conditions are for the purpose of biogas production and are therefore not those produced under typical management. Costigan (1993) reviews estimates for UK conditions and concludes that those of Safley and Westerman (1988) of 1.5 kg CH_4 per tonne of slurry per year are most appropriate. These estimates are used here which, given the

Table 6.3   Methane emissions per head from UK livestock

(1)	(2)	(3)	(4)	(5)	(6)	(7)	(8)
	Direct			Indirect			Total
	kg CH_4 year^{-1}	(weighted)	Slurry output tonnes per head per year	% housed	% stored slurry	Emissions (kg CH_4 year^{-1})	(kg CH_4 year^{-1})
Cattle							
beef	65.0 ⎫	55.12	2.4	50	20	1.42	56.54
cattle < 2 years	51.0 ⎭		5.5	75	20		
dairy in milk	95.0 ⎫	87.56	20.8	50	80	9.85	97.41
dairy not in milk	65.0 ⎭		12.4	50	20		
Pigs	1.5		1.3	100	75	1.43	2.94
Sheep	8.0		0.8–1.5	0	0	0	8.00

Source: Based on Costigan (1993) and own calculations.
Notes: Emission rate of CH_4 from slurry under UK conditions and typical management regimes is 1.5 kg CH_4 per tonne slurry per year (see text). For beef and dairy cattle the weighted emissions are based on the proportion in the UK herd for both direct emissions and from stored slurry (see Adger and Moran, 1993).

assumptions of the management regime (proportion of slurry stored and proportion housed in Table 6.3), equate to an average of 1.418 kg CH_4 per head per year for beef cattle, 9.850 kg CH_4 per head per year for dairy cattle and 1.429 kg CH_4 per head per year for emissions from stored slurry for pigs. Total emissions per head, calibrated for the conditions and profile of the UK herd, are presented in column 8 of Table 6.3.

Estimates of the monetary cost of global warming damage

The social benefits of greenhouse gas abatement are the avoided damage costs. Estimation of the optimal abatement which equates marginal abatement cost (primarily energy-based emissions) with marginal damage from warming, has formed the basis of several studies, resulting in ranges of cost per tonne of carbon (t C) at the optimal level (Ayres and Walter, 1991; Nordhaus, 1991; Cline, 1992). These studies are associated with various emissions scenarios and damage and abatement functions which rely critically on extrapolation from one economy (the US) across all regions of the world.

In Nordhaus (1991) the abatement function is designed to investigate an optimal strategy between the relevant greenhouse gases. Climate-induced sea level rise accounts for 91 percent of the damage as the agriculture and fisheries sectors, identified as the sectors of the economy most at risk, have low net impacts. The damage estimates are extrapolated to the global economy, implicitly assuming a similar sectoral profile for the world economy. This is far from accurate for developing countries where agriculture is a much more significant part of the economy, so economic and food security costs of climate change could be much higher. Nordhaus presents damage costs in US dollars per tonne of carbon equivalent for three scenarios which correspond to 0.25 percent, 1 percent and 2 percent of Gross World Income and equate to a range of US$8 per t C to US$66 per t C. Refinement of this approach was made by Cline (1992), who investigated long term impacts of atmospheric accumulation of trace gases over 300 years, and by Fankhauser (1992) who subdivided the climatic change impacts on the world economy by region.

Table 6.4 shows the range of estimates from a series of studies. The estimates of Ayres and Walter (1991) are greater than those of Nordhaus, though the same methodology was used. Ayres and Walter scale up Nordhaus' global warming damage and add further estimates of the likely consequences of sea level rise in terms of resettlement of displaced populations, to derive their higher estimate. The approach of both of these estimates (summing the damage costs and prevention costs of sea level rise impacts) can be questioned. The Ayres and Walter results may also be spurious according to Pearce (1992), due to an error in calculating the cost per tonne of emissions from their damage estimates.

The Nordhaus (1992) study (the DICE model) is more sophisticated. This is an optimal growth model extended to include a climate module and a damage sector which feeds climate changes back into the economy. The social cost values resulting from this are reported in Table 6.4. They are of a similar magnitude to his previous study, starting at US $5.3 per t C in 1995 rising to US$10.0 per t C in 2025. These and the Fankhauser (1994) results are current value estimates and relate to the time of the emissions. Fankhauser's (1994) estimates are based on a study which explicitly allows for uncertainty by incorporating random variables, such as the climate sensitivity to the doubling of atmospheric concentrations of CO_2 and the discount rate and damage estimates.

Choosing the discount rate for global warming analysis

There is a trade-off in choosing a discount rate in cost–benefit analysis that affects environmental assets. For renewable resources such as forestry the advantages of lower discount rates are that there is greater investment in longer term activities than would be the case if discount rates were high. However, the trade-off involved is that the scale and number of new investment projects will be increased

Table 6.4   Estimates of the social costs of CO_2 over time (US$ per t C)

Study	Period			
	1991–2000	2001–2010	2011–2020	2021–2030
Nordhaus (1991)			7.3	
Ayres and Walter (1991)			30–35	
Nordhaus (1992)	5.3	6.8	8.6	10.0
Fankhauser (1994)	20.4	22.9	25.4	27.8

with lower discount rates. But what is the appropriate rate when analysing global warming where the major impacts are cumulative and may not be experienced (e.g. in the case of sea level changes on defended coastlines) to any significant degree till some years in the future?

The analysis of global warming impacts is most sensitive to the discount rate used. The data below show the enormous range of a £1 (present value) impact occurring 200 years from now (say an impact at 2190 valued at 1990). The rates used in other studies of climate change impacts range from 0 to 3 percent; all well below what is considered to be a standard 'social' discount rate to be used by governments to appraise actions to be taken on behalf of society. Elements of the social discount rate include pure time preference, which reflects the observed patterns of economic behaviour that lead people simply to value welfare now rather than in the future, possibly because they assume that society will in general be better off in the future; and the opportunity cost of capital, which tends to be higher and reflects a real return on investing resources now and hence increasing welfare possibilities in the future. The social discount rate used in an analysis on behalf of 'society' is supposed to reflect both the opportunity cost of capital and the pure time preference rate, and hence enable public decisions to be taken with a greater emphasis on the long term than is possible for individual decision-makers. The resulting rate is a synthesis between these two rates; for the UK, for example, the standard rate used is six percent.

It is argued by many analysts of environmental problems such as global warming that lower rates than usually apply should be used for appraising situations where the costs and benefits are a long way in the future. The various rationales for discounting are questioned including the assertions that pure time preference is not relevant for public decision-making because pure time preference reflects only individual decisions; that continuous economic growth cannot be expected to continue to the future so the diminishing marginal utility of income factor is irrelevant; and that actual reinvestment and future welfare are not inextricably linked so the opportunity cost of capital factor is also not a sufficient rationale for discounting (Markandya and Pearce, 1991; Pearce and Warford, 1993; Price 1993). If strict intergenerational equity were required, a necessary sustainability constraint, then a zero rate would be required. As the appraisal of global warming requires a long, possibly hundreds of years horizon, the following figures show the impact of discounting at different rates on the resulting outcome:

Discount rate (percentage) 200 years from present	Present value of £1 impact
0.0	1.00
1.0	0.14
2.0	0.019
5.0	0.000 058

Cline (1992), for example, in his comprehensive study of climate change abatement and impacts, uses a rate of 2 percent. This is made up of an estimated social time preference rate of 1.5 percent, and a consumption equivalent measure of investment foregone in present costs, but these are weighted for whether present consumption or present capital investment would be used in global warming abatement. In other words, global warming abatement expenditure imposes production losses which affect the capital and consumption of present generations in proportion to the share of consumption (0.8) and investment (0.2) of total income in economies. The shadow price of capital for consumption equivalent results in a doubling of the real price of capital investment foregone. The social time preference rate of discount has the dual elements of pure time preference and a declining marginal utility of income. Thus Cline's rate is derived to be 2 percent. Even with this low rate, the Gross World Product, which Cline takes as the reference point for world impact of global warming analysis, is implicitly assumed to grow 25-fold over the 200 year horizon of his analysis.

Other studies derive rates based on similar methodologies, though debate exists over (a) whether a zero rate is appropriate, in the light of inter-generational equity and the likelihood that large-scale growth in economic welfare may be limited over the long run; and (b) the assumptions of the studies in deriving the social time preference rate (these issues are debated in Cline (1993) and Birdsall and Steer (1993)).

Table 6.5 shows the different rates used in three studies at the global level of aggregation of the impacts and abatement of global warming. All the studies use or derive low discount rates, the Fankhauser (1994) study deriving the rates from a similar methodology to Cline (1992), though using weights for capital and consumption equal to the relative global warming damage and having the rate vary between zero and 3 percent over time with changing income elasticity of utility.

The conclusions as to what the discount rate should be for the analysis of global warming seem to be: that the rate should be set low because the empirically derived long term rates give an indication of discounting in the range of 0–3 percent; that the estimation of costs and benefits is still important and is the relevant way to incorporate environmental assets into the cost–benefit analysis; and that the rate should be used consistently across all costs and benefits, rather than assigning different rates to environmental and non-environmental assets. A

Table 6.5   Discount rates and resulting shadow price of emissions from global warming studies

Study	Discount rate (%)	Shadow price of emissions ($ per t C)
Nordhaus (1991)	0–1	7.3–66
Cline (1992)	2	8.1
Fankhauser (1994)	0–3	20.4–27.8

'conservative' rate of 3 percent is used in the analysis presented here. This would seem to be a compromise between how other publicly funded actions are appraised in the UK (6 percent) and the school of thought advocating low discount rates specifically for the analysis of global warming.

Present values of benefits and costs compared

Subject to qualifications in deriving estimates of damage per tonne of carbon equivalents; on choosing the appropriate discount rate; and on using global warming potentials (GWPs) to compare across gases (see Chapter 1), the foregoing estimates are used to estimate a range of monetary values for CH_4 emissions from livestock under UK conditions. The results are shown in Table 6.6. The present value of benefits in £ per head is estimated, accounting for the damage values in US$ per tonne of C, the GWPs of CH_4 in relation to CO_2, and the exchange rate between US$ and £.

If the higher damage estimates and the global warming potential of direct and indirect effects of CH_4 in radiative forcing are taken (last two columns), it can be seen that the present value of the benefits of reducing CH_4 alone, is almost 25 percent of the cost of reducing agricultural support in the beef sector and almost 40 percent in the sheep sector. It is lower in the dairy cattle and pigs sector, because of the high present value of output in the dairy sector, and primarily because of the low price support (hence the market revenue reflects social cost) in the pig sector.

The environmental benefits of agricultural policy reform presented in Table 6.6 are a lower bound estimate, since they do not account for other negative environmental impacts of livestock such as nitrate run-off from grassland and its impact on human health and other aspects of water quality (Conway and Pretty, 1991). Agricultural policy reform in the EU is generally justified on the basis of budgetary cost and efficiency grounds rather than on environmental grounds; the latter are secondary objectives if accounted for at all. Greenhouse gas emissions have not been included in this calculus, and the analysis here has shown that this may form a significant part of the justification. Alternatively the conclusion to be drawn is that significant unaccounted for benefits accrue to society from reduction of the livestock herd.

6.4　AGRICULTURAL POLICY REFORM IMPACTS ON METHANE FROM LIVESTOCK

The estimates above allow an *ex ante* justification for reducing price support in the livestock sector: reductions are presently being brought about *despite* this justification, as a result of the other pressures within the EU described above. General price cuts of 15 percent were introduced in the beef regime in 1992 in Europe, with compensatory increases in premiums for beef producers to

Table 6.6 Costs and benefits of herd reduction compared

| | Present Value (costs) (£ per head) | PV of benefits (PV benefits as % of PV(C)) | | | | | |
| | | Low | | Medium | | High | |
		CH$_4$ GWP 11 Damage value US$8 t C (£ per head)	PV(B)/PV(C) (%)	CH$_4$ GWP 21 Damage value US$20 t C (£ per head)	PV(B)/PV(C) (%)	CH$_4$ GWP 63 Damage value US$66 t C (£ per head)	PV(B)/PV(C) (%)
Beef	4435	23	0.5	111	2.5	1 099	24.8
Dairy	10 687	40	0.4	191	1.8	1 893	17.7
Sheep	397	3	0.8	16	4.0	155	39.1
Pigs	1 823	1	0.1	5	0.3	57	3.1

Sources: Global Warming Potentials from (Shine et al., 1990; Isaksen et al., 1992).
Notes: Present value of costs (PV(C)) is the value of lost revenue to farmers of reducing herd size less the distortionary effects of public subsidy (see text and Table 6.2).
Present value of benefits (PV(B)) is the Nordhaus (1991) estimate (US$8 per t C) and Fankhauser's (1994) present value estimate (US$20 per t C). Present value of benefits is a range of values of global warming damage avoided (see text).
PV(B)/PV(C) represents the percentage of the costs of reducing farmers' income which can be offset by the global warming damage avoided alone.

maintain their incomes. The premium increases are, however, conditional on extensive production practices being adopted, including specific stocking rates in terms of animals per hectare. The effects of changed agricultural support in the livestock sectors in terms of aggregate livestock numbers are difficult to predict. The dairy sector is subject to quotas, which successfully limited herd sizes and output from the mid-1980s onwards with the resulting aggregate emissions for the past decade shown in Figure 6.1. The reduction in dairy cattle numbers over the period is reflected in the gradual decline in emissions. Pig and sheep emissions are aggregated, with the increasing trend attributed largely to rising sheep numbers.

Aggregate emissions and other pollution externalities are dependent on both total numbers and the management regime. If the livestock sector is to be part of a greenhouse gas emission reduction strategy, then the extra costs of emission reduction need to be compared to alternative emission reduction strategies. However, the present reform of the CAP in the EU does not encourage afforestation and other forms of emission reduction as shall be seen in the following chapter. Although the benefits accrue to society from abatement of CH_4 emissions, the extent to which they *are* abated depends on *private* decisions affected by market structure, regulation and fiscal incentives, and these are culturally and institutionally bound.

Intuitively, it would seem cost-effective to make some emission reduction in the agricultural sector, both for the reasons outlined above concerning the costs of agricultural support in many industrialised countries and regions, and because of the propensity of the high potency gases such as CH_4 and N_2O emitted from agricultural practices. The sections above have outlined how economists estimate the benefit to society of abating CH_4 emissions, but the extent to which they are abated depends on private decisions affected by market structure, regulation and fiscal incentives. Although these issues have been explored in depth for the energy sector in relation to greenhouse gas emissions, few estimates exist of the cost of abatement of the multiple sources of gases from agriculture, with the exception of the afforestation to sequester carbon option.

One such study which estimates the costs of abatement through changing agricultural practices is that of Adams et al. (1992). The principal methods of reducing CH_4 emissions from livestock in the short run are to change the feed regime to make it less CH_4 producing, but at a cost of animal productivity; and to capture or reduce emissions from the anaerobic lagoons of slurry. Technical solutions such as covered lagoon systems and tank digesters which use the CH_4 to produce energy for on-site use are common in the US, but have high initial capital cost (Gibbs and Woodbury, 1993). The simulated demand change (reduction of quantity demanded at each price) could be as a result of changes in the price of substitutes or exogenous change in tastes, such as increases in vegetarianism.

Alternatively, this could be seen as the result of the imposition of a tax on meat products because of their greenhouse gas externalities. This would be a modest

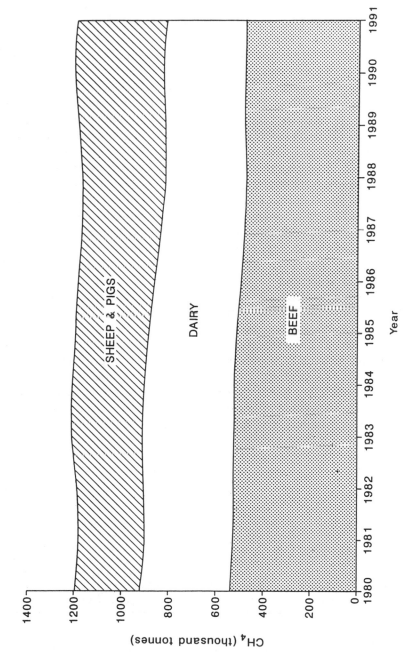

Figure 6.1 Methane emissions from UK livestock, 1980–1991. Source: MAFF (various years) and author's calculations

tax in comparison with proposed or already existing taxes on oil and products, which range from 30 percent of producer price in the US to over 60 percent in the UK and some European countries (Poterba, 1993). Coal, an energy source with even higher greenhouse gas emissions ratios than oil, is presently subsidised in many countries; the UK subsidises coal to the equivalent of US$46 per tonne of carbon emissions and Germany by US$77 per t C (reported in Poterba, 1993). A tax on CH_4 emissions would be the direct method of reducing emissions and creating incentives for utilising the escaping gases, but it should be noted that similar to coal, a tax on agricultural products for their greenhouse gas externality would be a *net reduction of subsidy* in countries such as in the EU and the US, rather than net taxation.

To reduce CH_4 from rice production in the US, where large amounts of fertiliser are used which enhance the methane to rice production ratio, Adams et al. (1992) estimate the cost of reducing the fertiliser input. Finally an estimate is made of reducing N_2O emissions by reducing nitrogen-based fertiliser application to arable land, again with consequences for productivity. The impact of fertiliser reduction on economic welfare comes about through the simulation of the reduction of anhydrous ammonia, which brings about the release of large amounts of N_2O compared to other nitrogenous fertilisers. A 50 percent reduction in anhydrous fertiliser use in the US would reduce yield across the range of crops grown by 3–15 percent.

The results of a simulation of these various options for the US are given in Table 6.7. The results are changes in economic welfare brought about by the various options. Economic welfare is the net impact of price changes and quantity changes on consumers and producers. If the price of agricultural produce rises, and this is passed to the consumer (the extent being determined by the responsiveness of demand to price) then the consumer is made worse off, all things else being constant. Rising prices are good for the producers, farmers in

Table 6.7   Estimated abatement cost of reducing CH_4 and N_2O from US agriculture

Strategy	Cost per tonne (US$ per tonne)	Cost per tonne CO_2-equivalent (US$ per t CO_2-equiv.)
Livestock feeding regime inducing 5% yield reduction	$1166 per t CH_4	56–106
5% reduced demand for beef (equivalent to a tax)	$4180 per t CH_4	199–380
Reduced fertiliser application to rice by 50%	$590 per t CH_4	28–54
Reduced fertiliser application to rice by 100%	$663 per t CH_4	32–60
Reduce fertiliser applications to arable crops	$4708 per t N_2O	16–17

Source: Adapted from Adams et al. (1992).
Notes: Costs per CO_2-equivalent are based on converting CH_4 and N_2O by global warming potentials. The range given reflects two GWPs: 100 year direct only GWP ($CH_4 = 11$; $N_2O = 270$) and 100 year estimated direct and indirect GWP ($CH_4 = 21$; $N_2O = 290$) (see Chapter 1).

this case, so the benefit to the farmers must be held against the disbenefit to the consumers of rising agricultural prices. The quantities of food produced and bought also change with changing relative prices, so these impacts also have to be accounted for. The impact of reducing fertiliser application on arable crops, for example, leads to higher prices, and given the elasticities of demand and supply of the products, Adams et al. (1992) estimate that the cost would be borne exclusively by consumers through higher prices. The net impact of all these changes is the economic welfare effect. The measure does not discriminate between who receives the benefit or the disbenefit; it simply measures overall change and implicitly assumes that distributional issues of the policy changes should be dealt with separately.

The results show the estimated costs from the simulated scenarios, with the costs of reducing CH_4 from rice being much lower than from the livestock sector. The costs of N_2O abatement are high per unit weight of N_2O, but this has a much greater impact on the greenhouse effect. The last column shows the cost as CO_2-equivalent. On this basis, abatement strategies to reduce N_2O emissions are very attractive. Two issues need to be raised here.

First, the distribution of costs and benefits from abating emissions in agriculture requires attention. As explained above, the costs of agricultural support are shared between consumers and taxpayers and changes in agricultural production designed to reduce greenhouse gas emissions may be at odds with other policy objectives. If these abatement strategies are implemented, compensatory measures within the food and agricultural sectors may have to be undertaken. The ability to compensate and redistribute income is the single most important factor making a tax on carbon emissions appealing to government, even if it is narrowly focused on the energy sector and thereby not optimising emission reduction across sectors. The potential impacts of carbon taxes on income distribution are significant (see Rosenberg and Scott (1994) on the impacts of energy taxes on agriculture), though the impact of policy changes designed to reduce greenhouse gas emissions associated with agriculture have not been researched.

Secondly, the scope for abatement from agriculture is ultimately limited, a constant theme throughout this volume. The results presented are applicable only to the US system of agricultural production and prices, and the costs are likely to be significantly different in other regions of the world. The demand for food in general rises constantly with human population and strategies which reduce yield are short term solutions only: emissions reduction must eventually be achieved in agriculture, as in energy and other sectors, through reducing emissions–output ratios by substituting new technologies. Further, as indicated in Chapter 1, even if all emissions associated with agriculture were reduced to zero, this would fall short of goals already deemed desirable under the Climate Change Convention. For example, the costs of abatement in rice production for the US are modest but the US constitutes only 1 percent of the world total rice production, and feasible alternatives for emission reduction for the majority of rice-producing areas are significantly constrained by population pressure and lack of desirable substitute products, as discussed below.

6.5 STRATEGIES FOR REDUCING METHANE EMISSIONS FROM RICE PRODUCTION

The other major source of CH_4 associated with agricultural practices is that associated with rice growing. Chapter 4 showed that the emissions of CH_4 from rice production are concentrated in Asia. Total global emissions are approximately 95 mt CH_4 year^{-1}. This section gives the context of world rice production and discusses options for reducing emissions per unit of food produced.

Given the estimates of global methane emissions and the predominance of those from relatively low-income countries, mitigation strategies for reducing emissions from paddy rice would seem to be limited. Future emissions will be determined both by the fluxes per unit of rice production and the total quantity of rice produced. Rice demand is determined by population growth in the rice-producing regions, assuming that nutrition patterns do not change radically. Rice production is a function of the land and water available, the rice variety and the farmer's agronomic practices.

In the major rice-growing areas there are large discrepancies in both agricultural productivity and in the levels of economic development. This is illustrated in Table 6.8 where average agricultural productivity across all arable land uses varies in Asia from 2.0 t ha^{-1} for Nepal, to 13.3 t ha^{-1} for Japan. Although these large discrepancies exist in productivity, countries of Asia have in common the extent to which the area of agricultural production is limited, approximately 90 percent of the physically cultivatable area being already cropped (see Table 1.8). An agricultural policy objective of self-sufficiency is common to most countries of the world and the Asian rice-growing economies of Japan, Korea and Taiwan restrict the conversion of land to non-agricultural use. As each nation's ability to increase agricultural production is dependent on land availability and on productivity, the intensive production of food, and of rice in the Asian context, is being brought about through the increased use of irrigation and of fertilisers (see Table 6.8). Both of these have implications for CH_4 emissions from rice production.

The growth in population and in the area devoted to rice production in Asia (which represents over 85 percent of the world rice area) over the last 20 years is shown in Figure 6.2, with population projections into the next century (FAO, 1992). It is expected that rice production will increase by between 50 and 100 percent in the next three decades (Neue, 1993). It is evident that the growth in population is outstripping the increase in land area, and that most of any increase in rice production will only be achieved by more intensive agriculture, especially the further development of irrigation and the introduction of even higher performing rice varieties. This has been the basis of the significant rice production

Table 6.8 Changes in agricultural productivity and fertiliser use in 1980s for major rice growing economies of Asia

	Growth in agricultural production, 1980–1988 (%)	Agricultural productivity per unit area, 1988* (t ha^{-1})	Growth in productivity, 1980s (%)	Fertiliser use, 1988 (kg ha^{-1})	Growth in fertiliser use per unit area 1980s (%)
Bangladesh	8	2.95	5	58	61
India	32	3.17	13	65	41
Nepal	23	2.02	−4	16	23
Pakistan	45	4.13	17	106	43
Sri Lanka	2	5.50	−4	159	37
Indonesia	37	4.73	6	95	40
Malaysia	50	3.00	17	160	57
Philippines	9	4.22	−16	45	36
Thailand	20	4.29	−4	29	53
China	43	5.53	12	171	44
Japan	0	13.30	4	565	6
Korea	10	9.67	11	404	53

Source: Compiled from Ahmed (1993) based on FAO data.
*Agricultural productivity is averaged over rice crop production.

184

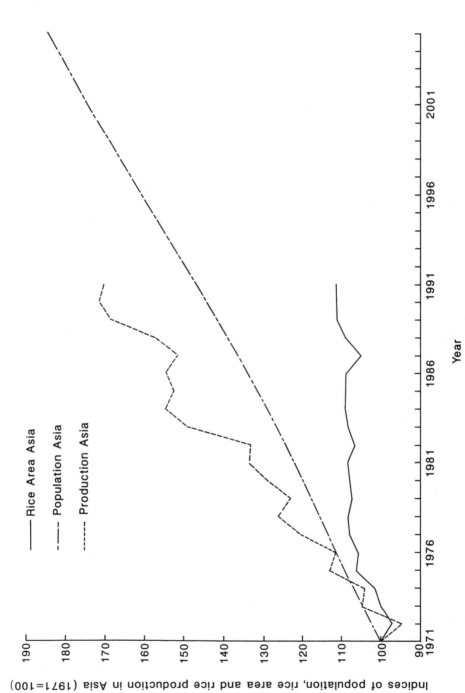

Figure 6.2 Historic and projected rice production, rice area and projected population for Asia

increase in the last 20 years. The interplay of the present world population, the availability of land and the proportions of land that are considered cultivable and irrigable are discussed by the FAO (1991). Their future projections of the ability of the land resource to support the population are pessimistic for many countries as over 90 percent of the estimated potentially cultivable area is already in use in Asia (compared with less than 50 percent in Central America and less than 25 percent in Africa and South America).

Further uncertainty concerning future rice production and hence methane emissions is from the feedback from climate change itself. The impacts of the emissions of the greenhouse gases will be felt directly on agriculture. The agronomy of all crops will be affected by both temperature and precipitation changes and by the increased atmospheric concentration of carbon dioxide. For rice, increased yield due to CO_2 fertilisation may occur at concentrations higher than at present (around 355 ppmv) which will cause $1-2\,°C$ global mean warming, but Rosenzweig et al. (1993) estimate that the net yield increase turns negative as temperature increases by $3-4\,°C$. However, these crop model predictions hold precipitation constant and it is the seasonal water availability which most heavily influences rice yield. The feedback impacts of climate change on rice production are therefore highly uncertain.

Mitigation strategies to reduce methane emissions must intervene by reducing the methane flux per unit rice production. If they are to be successful, they must bring some advantage to the individual farmer: an increased net benefit in value of production per unit of inputs, be they physical inputs such as mineral fertilisers, or labour inputs; a better deployment of inputs, or a reduction in risk.

At the farm level some intervention can take place. Micro-irrigation and the availability of some inputs can be controlled to a significant extent by the farmer. Organic matter generated within the farm and used as fertiliser increases the methane flux, and the depth of water in the field can be controlled within certain limits by opening and closing sluices or bunds. Subject to these influences, the paddy rice farmer will choose which rotations of which cultivars of which crops to grow, and at what planting density; paddy rice will usually be the principal crop within the rotation. High yielding varieties grown according to recognised 'best' practices tend to produce lower methane emissions per unit, because of their different fertiliser demands (a higher proportion of mineral nitrogen) and their shorter growing periods.

The macro-irrigation of the land is usually beyond the influence of the individual farmer, but the micro-irrigation can be controlled within limits. The degree to which the soil is reducing is determined by the soil type, the land preparation and the water management. The soil type cannot be radically changed by the farmer, although his techniques of land preparation may modify it. Repeated ploughing and puddling of the soil during land preparation tends to reduce permeability, and intensify the reducing conditions when excess water is applied. The methane flux is linearly related to the depth of water up to a maximum at about 10 cm in cold climate wetlands, so that drainage where possible to this depth or below should reduce emissions from paddy fields (Sebacher et al., 1986; Wassmann et al., 1993). This will allow economies in irrigation water where this is scarce.

Draining the land to field capacity or letting it dry out further through evapo-transpiration or percolation to groundwater allows atmospheric oxygen to re-establish oxidising conditions from the soil surface downwards. This reduces methane production. Farmers usually partially drain their land for weeding and applying fertiliser, and allow it to become progressively drier towards harvest. If periodic draining is encouraged the oxidation reduces the phytotoxic effects of the intermediate products of anaerobic decomposition, and also potential rice yield increases in the range 10–20 percent (Cheng, 1984). Removal of these phytotoxins by oxidation following drying by evapo-transpiration may be preferred by the farmer to their removal by percolation as it avoids the significant simultaneous loss of soil nutrients to groundwater (Ponnamperuma, 1984); percolation is, however, more effective in reducing methane emissions by removing methanogenic substrates, particularly in cold, saline and alkaline soils (Wassmann et al., 1993). The organic substrates for soil methanogenesis have two principal sources: the organic matter that is applied as fertiliser, and root exudates from the rice plants. Different cultivars exude different quantities of reducible organic compounds into the soil, and the composition of the exudate also varies from one cultivar to another. Very little experimental work has been carried out in this area, but the selection and development of new rice varieties have the potential to reduce substantially the associated methane fluxes (Wassmann et al., 1993).

A further option for reducing CH_4 emissions is in the handling of rice straw and waste. At higher temperatures, the proportion of methane in the mixture of carbon dioxide and methane that is produced from anaerobic straw decomposition increases. Thus global methane emissions would be reduced if farmers in warmer climates or at warmer periods of the year allowed the straw to decompose aerobically before incorporation into the soil.

The mitigation options which increase the production to methane flux ratio also have impacts on other greenhouse gases (summarised in Table 6.9). Although the use of inorganic fertilisers and improved water management will reduce methane fluxes, straw and organic fertiliser may increase fluxes. Field results reported by Denier van der Gon et al. (1992) suggest that the mode of fertiliser application is unimportant and is unlikely to offer feasible mitigation options. As can be seen from Table 6.9, however, the opposite impact on emissions of N_2O tends to occur, and the CO_2 production from increased management results from the mitigation of methane options identified. These conflicting impacts remain to be quantified; a necessary step to appraising the likely net greenhouse effect impact of future rice growing.

Table 6.9   Greenhouse gas impact of wetland rice management strategies

Improved management through:	CH_4	N_2O	CO_2
Use of inorganic fertilisers	−	+	0
Addition of straw	+	−	+
Addition of compost	+	−	+
Water management improvement	−	+	+

Source: Based on Bouwman (1990c, p. 10).
Note: (+) is increased emission and (−) is decreased emission.

6.6   CONCLUSION

The abatement of greenhouse gases potentially has significant influence on the whole direction of the world economy in the next century. If abatement is brought about by a tax on carbon emissions, or more likely by a tax on energy, it could revolutionise the basis of public finance and government revenue throughout the world (Barker, 1993), especially to the extent that the economy is locked into energy consumption patterns and the resulting inelasticity of demand for energy. The introduction of fiscal or other measures to reduce emissions requires much prior knowledge if the mechanism is to approach a cost-effective solution. First the 'social cost' of emissions must be estimated. The optimal taxation rate to meet the narrow economic definition of efficiency is equal to the cost to society of emitting the greenhouse gases. In this chapter estimates of these costs and the impact of global warming on future generations have been reviewed. Economists cannot produce reliable forecasts concerning the economic state of the world 50 years from the present, and to this uncertainty is added the uncertainty surrounding climate change impacts.

The estimate of the economic value of these impacts clearly cannot take into account many of those impacts which are either difficult to quantify in economic terms (such as the costs of adaptation to climate change) or are outside the scope of economic analysis. Nevertheless the ranges of values given in this chapter (best guess US$20 per t C global damage) are useful lower-bound values to society of reducing emissions.

The reduction of CH_4 emissions from livestock is an unpriced, and generally unaccounted for, benefit of agricultural policy reform in highly subsidised agricultural sectors. Given present and projected future prices in the livestock sectors in the UK, it is estimated that up to 40 percent of the cost of reducing support in the sheep sector, for example, will be offset simply by the benefits accruing to society from reduced CH_4 emissions. These estimates are based on the US$20 per t C global damage estimates, which as discussed are a lower-bound estimate of monetary value of avoiding the impacts of global warming. Further, since the Climate Change Convention has come into force, the UK and the other signatories are committed to reducing their emissions of all greenhouse gases. Reduction of CH_4 or of N_2O emissions could form part of this strategy. Whether this is desirable compared to the other options available, such as afforestation or energy efficiency, depends on their relative cost.

The second area in which knowledge is required for economic policy is on potential abatement costs. Very little research has been carried out into the costs of abatement of non-CO_2 gases, with a disproportionate share being focused on energy sector modelling. The research reported here shows that there are feasible and cost-effective opportunities for reducing agricultural emissions, given the present structure of economic support for agriculture in the US and elsewhere. However, the scope for abatement in agricultural practices is ultimately limited on the global scale by the size of the contribution of the agricultural sector and by the pressures on agricultural resources and technical progress to produce food and fibre in the future given human population growth. As Poterba (1993) notes, the removal of energy subsidies in various nations would produce short term reductions in CO_2 emissions from that sector through the price effect on consumption. Similarly, this chapter has illustrated the fact that agricultural production is subsidised in many industrialised countries, and the *real cost* of emission reduction needs to take into account the distortionary signals of agricultural policy. There may be many 'no-regrets' opportunities in the agricultural sector.

CHAPTER 7

Forestry Options for Offsetting Emissions

7.1 INTRODUCTION

Deforestation is the single most important cause of the greenhouse effect related to land use. Stemming deforestation is an important policy measure to reduce emissions. The enhancement of sinks through *afforestation*, however, is not necessarily a feasible, equitable or cost-effective strategy for reducing aggregate emissions or recovering the lost biomass carbon. In terms of timing, halting deforestation brings about immediate benefits whereas the enhancement of carbon sinks through afforestation has its effect on the global carbon cycle over a longer time frame. The rate at which carbon sink enhancement takes place has therefore to be compared with its alternatives. These alternatives include halting deforestation but also the non-land use related actions such as the reduction of emissions from the energy sector. Both of these alternatives are more appealing by bringing about instantaneous reduction of net emissions of greenhouse gases (see Price and Willis (1993) on the economics of this time factor). This chapter explores the cost aspects of afforestation in a UK context, showing that even if afforestation is desirable, distortionary agricultural and land use policies militate against this option. We use the term 'afforestation' to mean a range of options for forest sink enhancement. Afforestation therefore includes natural regrowth or reforestation of forests, agroforestry and multiple land uses which include trees, as well as the establishment of new forest areas (the common meaning of the term).

The issue of carbon sequestration and storage in biomass can be analysed as a stock–flow problem. The present *stock* of carbon is being reduced through deforestation, but the stock is ultimately constrained through land availability and other factors. One of the major unknown factors is the impact of CO_2 fertilisation and of climate change itself on forests. The *flow*, or potential accumulation rate, is equivalent to the abatement of emissions. Sink enhancement and emissions reduction for all purposes result in the same outcome: the removal of greenhouse gas from the atmosphere.

This chapter first illustrates the technical feasibility of afforestation to offset the greenhouse effect, by reference to various studies of land availability. These have tended to concentrate on the tropics, as this has been the site of greatest recent deforestation and hence assumed greatest availablity for afforestation. The results show that a significant proportion of the world emissions could be offset by afforestation either in the tropics, or even in temperate regions. The costs of afforestation as a strategy to reduce emissions are then discussed for various regions in the world, based on national or cross-national studies. Afforestation may appear to be cheaper in tropical countries because the market cost of agricultural land is assumed to be low in many developing countries, compared to agricultural land in the US and Europe. The phenomenon of higher land prices is, however, simply an illustration of how political and economic support for agriculture distorts the markets related to land use activities and hence distorts land use decisions themselves. In reality, large-scale radical land use change through afforestation in the tropics would have many social and economic consequences.

The chapter then considers the context of afforestation in the UK. The UK along with many other industrialised nations has passed the forest transition (Mather, 1990) so that the proportion of forest land to total area is increasing. Nevertheless forestry has a high *financial* cost because of the opportunity cost of land, brought about by the high levels of price support for the agricultural sector. The *true* cost of afforestation may well be lower than the estimated financial costs. The system of agricultural support has been under pressure because of its high costs, and enhancement of the environment in general has been recognised as a legitimate objective of the reform of agricultural policy. However, the mechanisms of agricultural support do not allow this potential environmental enhancement including afforestation, habitat restoration and other measures, to be realised.

7.2   AVAILABILITY OF LAND IN THE WORLD FOR AFFORESTATION

The physical approach to land availability for afforestation can be conceptualised in many ways: estimates could be made of the maximum sequestration if all land in the world were afforested with the fastest growing vegetation; if all land that had once been forested were returned to the same forest type; or if all land not necessary for agricultural production were afforested with tree species indigenous to the region, or with fast-growing exotic species. Only in the last of these (afforesting of surplus agricultural land) is there consideration of alternative land uses and the trade-offs or opportunity costs involved. The opportunity costs of afforestation, wherever it occurs, are dependent on the demand for alternative uses, predominantly agriculture. This section illustrates the results of physical assessments of global reforestation to offset greenhouse emissions, before discussing how an economic appproach assists in deciding between abatement options, and highlights the dangers inherent in single-objective strategies.

One relevant study considers the potential for carbon sequestration through afforestation (Houghton et al., 1993). Their criterion of land availability is that all land which was previously under forest prior to human disturbances is technically feasible for afforestation. If all this land were managed to maximise carbon sequestration, 160–170 bt C could be accumulated. This is equivalent to the accumulation of carbon in the atmosphere since the start of the industrial revolution or about 25 years of fossil fuel emissions at present rates. The results of this assessment are surprisingly optimistic, all the more so because the study was limited to the analysis of the tropics and sub-tropics (35 °N–35 °S).

The emphasis on the tropics in estimating the physical availability for carbon sequestration in many studies occurs for a number of reasons:

— the tropics are the site of the most recent large-scale deforestation, whereas temperate areas have been in alternative land use for centuries;
— tropical forestry has higher potential growth rates than temperate forests, allowing quicker carbon sequestration;
— the costs of forest establishment are lower in the tropics because of lower land and labour costs;
— demand for forest products, such as fuelwood and other services, is greater in the tropics.

The availability of land for increased carbon sequestration was assessed by Houghton et al. (1993) by comparing present land use cover in the study area from satellite imagery, with that before human disturbance (based on the assessment of Matthews, 1983). The resulting areas and the estimated sequestration potential are shown in Table 7.1. The carbon sequestration rates are based on the plantation management of those areas presently logged, and on the increase in biomass carbon in grassland areas through agroforestry. The biomass for plantation forests was assumed to be half of the maximum biomass of mature stands, as they are harvested on a rotational basis.

Given the assumptions of the management regimes, Table 7.1 shows that agroforestry is the single largest potential sequestration option (72 percent of the total), due to both the high potential sequestration rate per unit area and the large area. The agroforestry potential areas were predominantly in Africa and tropical America (85 percent of the total), with only 15 percent in Asia. Protection of secondary forests was estimated as the single most important option in Asia with the overall geographic spread being: Africa, 40 percent; Latin America, 39 percent; and tropical Asia, 21 percent.

Houghton et al.'s results do not claim to prescribe how the afforestation of available land should come about or the likely availability of land. A next iteration would be to factor in human population change to show decline in available land area. Such an approach is adopted by Bekkering (1992), again

Table 7.1   Physical assessment of carbon sequestration potential in the tropics

	Area in tropical and sub-tropical continents available			Average carbon increment (t C ha^{-1})	Total potential accumulation (bt C)
	America (million ha)	Africa (million ha)	Asia (million ha)		
Plantation					
forests	33	171	26	61–78	
woodlands	11	7	3	29–34	
Total	44	178	29		17.8
Protection					
forests	575	26	336	17–44	
woodlands	22	0	0	26–30	
grasslands	37	0	57	5	
Total	634	26	393		28.6
Agroforestry				59–69	
forests	355	241	194		
woodlands	142	392	44		
grasslands	240	256	32		
Total	737	888	270		120.4
Total	1415	1093	692		166.8

Source: Adapted from Houghton et al. (1993).
Notes: Carbon increment based on (standing biomass and soil carbon for undisturbed forest, grassland etc.) - (estimated present carbon in biomass). Ranges given for carbon increments are the area-weighted average of the increments across the three tropical continents in the assessment. Areas based on: (areas prior to human disturbance) — (present satellite imagery of land cover). Tropical and sub-tropical regions include most areas 35 °N - 35 °S with the exception of North Africa, Australia, southern China and middle east (see Houghton et al., 1993).

with the intention of estimating potential carbon sequestration from tropical reforestation. However, Bekkering simply assumes that land availability is a function of the total land area and carrying capacity following the simple relationship:

$$\text{Area available for forestry in year } t = \text{total unforested area in year } t - \text{area required for agriculture in year } t$$

and

Area required for agriculture in year t = population in year t / carrying capacity

This approach to estimating land availability, although taking on board population aspects of future land use, relies exclusively on the *carrying capacity* concept to incorporate all the necessary social forces driving land use. To estimate carrying capacity requires knowledge of not only the physical attributes of land and the level of technology which combine to determine the productivity of land, but also knowledge of the land requirement, which is influenced by urbanisation and non-agricultural income of land users. The demand for land is also fundamentally determined by localised institutions and cultural norms, such as inheritance law and practice. This combination of criticisms of the carrying capacity approach leaves the physical assessment of carbon sequestration potential as only a first step in policy design or appraisal.

7.3 ECONOMICS OF AFFORESTATION

The two studies highlighted above approach the problem of afforestation by identifying maximum abatement strategies for the feasible land identified: Houghton et al. (1993) estimate that 25 years of current fossil fuel emissions could be offset, while Bekkering (1992) identifies the countries where afforestation could best take place, given the requirement for agricultural land. Alternatively an estimate of the *minimum* area required could be estimated to reach a particular goal: in the case of Sedjo and Solomon (1989), this is to estimate what area is required to remove 2.9 bt C which they take as being added annually to the atmosphere. This target results in a need for 465 million ha of new afforestation, given an average yield class of $15 \, m^3 \, ha^{-1} \, year^{-1}$ ($= 6.24 \, t \, C \, ha^{-1} \, year^{-1}$). The costs of establishing forests on this magnitude in the United States are estimated as the sum of the actual costs incurred in planting the trees and the cost of purchasing the land. For the US this is estimated as:

Total costs = (establishment costs per hectare + land costs per hectare) * area required

and

Area required = target sequestration/sequestration rate per hectare

For temperate forestry in the US:

Area required = 2.9 bt C/ $6.24 \, t \, C \, ha^{-1}$ = 465 million ha

Costs = US$(230 + 400) × 465 million ha = US$372 billion

Sedjo and Solomon (1989) give this figure of US$372 billion, equivalent to 8 percent of the gross national product of the US, as a lower-bound estimate and purely as a financial cost rather than a change in economic welfare. The cost is likely to be higher because the scale of forestry required would mean bringing productive agricultural land into forestry, with accompanying much higher land costs. Further, the real cost of forestry includes the opportunity cost: the cost of the next best foregone alternative. Although agricultural production is higher than it would be under free market conditions due to distorting price support in both the US and in Europe, the costs of agricultural production foregone are not included in the above calculation, and hence the figure represents a lower-bound estimate.

For tropical areas, Sedjo and Solomon (1989) also make a first attempt to estimate the costs of stabilising atmospheric CO_2 concentrations through forest planting. Again this is based on the 2.9 bt C added annually to the atmosphere, and assumes a 15 m^3 ha^{-1} $year^{-1}$ forest growth rate. The costs of establishment are estimated as US\$400 ha^{-1} and the cost of land purchase at zero. The aggregate costs are therefore:

$$Area\ required\ =\ 2.9\ bt\ C/6.24\ t\ C\ ha^{-1}\ =\ 465\ million\ hectares$$

$$Costs\ =\ US\$(400\ +\ 0)\ \times\ 465\ million\ hectares\ =\ US\$186\ billion$$

The costs are significantly lower for the tropics than for the US for the same carbon sequestration target primarily because of the assumed zero land cost. This assumption is that land is effectively 'surplus' and has no opportunity cost. Yet this estimate for tropical zones must also be regarded as a lower-bound estimate as any forestry activity, even if restricted to the planting of shelterbelts along the edges of agricultural land, has costs in terms of lost area for other purposes (Anderson, 1987). A range of forestry options has been considered by Dixon et al. (1993a), in an analysis of both costs and carbon sequestration rates across a range of countries. Their results on the average cost of sequestration, again based on establishment costs only for a range of tropical and temperate countries, are shown in Table 7.2. This shows a large range in estimated costs. However, this tells us little about whether forestry options are likely to be taken up by land users. Laws and regulations vary across countries, and in many cases the rights to own, manage and harvest timber and other forest products are often not vested with forest users themselves. All these factors will affect decisions about land use and afforestation.

The real cost of afforestation options should then be the establishment cost and the opportunity cost of land plus any positive or negative environmental impacts associated with forestry. The opportunity cost of land is not necessarily the market price for land, as the market price is distorted by government expenditure. In the case of countries where assets which are internationally tradeable are involved in the plantation and as part of the opportunity costs of the land, the value of the foreign exchange component should be modified to reflect the true scarcity of these resources in the economy (e.g. ODA, 1988). However, the data requirements for this economic analysis are formidable, and hence most assessments of abatement costs remain at the 'financial' level rather than the 'social economic' level. Further analysis which takes account of these factors, as well as legal and property rights issues, is therefore necessary to gauge more accurately the potential of afforestation schemes.

Table 7.2   Estimates of the cost of carbon sequestration through afforestation options in selected countries

Country	Estimated sequestration rate (t C ha^{-1})	Average cost of sequestration (US$ per t C)
Temperate		
Argentina	65.0	25.0
Canada	40.7	11.5
Chile	75.9	2.7
Finland	40.9	27.3
France	115.9	16.6
Germany	56.0	32.4
New Zealand	94.0	52.4
Nepal	78.6	0.5
Pakistan	98.8	2.5
South Africa	110.0	8.7
USA	77.0	5.5
Former USSR	25.0	4.6
Tropical and sub-tropical		
Australia	107.0	5.9
Brazil	63.5	27.4
Burkino Faso	92.8	13.3
China	76.3	5.2
Columbia	56.7	20.6
Congo	111.0	3.0
Costa Rica	46.3	31.0
Côte d'Ivoire	60.2	34.1
Ecuador	67.2	8.7
Egypt	36.5	77.9
Ghana	129.9	6.0
India	78.0	26.8
Indonesia	99.0	6.8
Madagascar	75.0	1.8
Malaysia	84.0	3.5
Mexico	99.0	4.5
Philippines	44.6	2.7
Senegal	128.8	2.9
Thailand	79.2	1.7
Togo	167.0	4.6
Venezuela	38.9	50.7
Zaire	59.5	41.7

Source: Dixon et al. (1993a).

Notes: Carbon sequestration rates based on a range of management options including natural regeneration and plantation establishment. Reported costs are costs of establishment including labour costs, transportation and infrastructure, recurring over a 50 year period, discounted at 5 percent. Land purchase or opportunity cost of land are not included. Temperate and tropical and sub-tropical categories overlap in some countries.

7.4 AFFORESTATION OPTIONS IN THE UK

Set-aside land

As outlined above, the cost-effectiveness of carbon sequestration through afforestation is dependent on the opportunity cost, i.e. on the demand for land as a productive resource. In the countries of the European Union (EU), the primary demand for land for agricultural purposes has in effect been declining in the post-war period as price support mechanisms encourage greater productivity through increased capitalisation of agriculture and rapid technological advance. Indeed the European Union has become more than self-sufficient in all temperate foodstuffs, and in the case of some traditionally tropical products such as sugar, for example, where previously Europe imported cane sugar from tropical countries, it has now substituted domestically grown sugar beet. Where the EU produces more than its domestic requirements, the surpluses have been exported with the assistance of subsidies, depressing and destabilising world market prices.

The food security objective of agricultural policy (as enshrined in the Treaty of Rome which set out the framework of the Common Agricultural Policy) has thus been fulfilled. The demand for land for agricultural production has therefore been decreasing in social terms, though the price of agricultural land has not decreased in real terms as the mechanism of price support has, until recent years, maintained the returns to land. Various estimates have been made of the 'surplus' land in the UK, which may become available based on technological advance and the quantity of production demanded. These estimates show a range of 'surplus' land in the UK of between 0.7 and 6 million hectares becoming surplus *to agriculture* by 2000 or 2005 (Harvey, 1991). This surplus is defined by reference to projected domestic consumption within the EU, and by taking average yields per hectare for the various commodities; this is then translated into an area requirement and hence to a surplus area estimate. However, as Harvey (1991) has pointed out, if these predictions were converted into annual rates, this would be by far the fastest land use change in any historical period in the UK. The total change in agricultural land in the 20 years since the UK joined the Common Agricultural Policy (1972 – 1992) was a net change out of agriculture of only 0.7 million ha, and a large proportion of this has been to forestry (Adger, 1993). Even with restructuring of the agricultural sector towards more capital-intensive agriculture, future responses will likely be to release capital and labour rather than land, to produce the reduced demand (Harvey, 1991). Further, in almost every country in the world there are land use restrictions primarily in the form of regulation to prevent the change of agricultural land to non-agricultural uses. In most industrialised and land-scarce countries, the demand for non-marketed amenity goods is high and rises with rising incomes (Walsh, 1986). Amenity demands for non-built land also ensure that land for offsetting the greenhouse effect will always be scarce, especially at the scale required to bring about major carbon sequestration.

In summary, in circumstances where food production is a central objective of economic and social policy, the opportunity cost of land is high. In high-income countries where intensive input agriculture has fulfilled the food security objective, the opportunity cost of land is high for other reasons such as increasing demands for amenity. However, the present agricultural price support mechanism in the EU does have undesirable impacts, principally the monetary cost to consumers and taxpayers. To rectify this situation, a set of policies has been pursued which does indeed create land available for other purposes than agricultural production. The reality of land surplus to agriculture has therefore been seen in the recent reform of the Common Agricultural Policy which, as well as making the price changes for beef and other livestock described in Chapter 6, also introduced policies to set-aside land from agricultural production. Set-aside, or land diversion as it is known in the US, involves the payment of subsidies to farmers to take land out of agricultural production. In the case of the recent EU schemes it involves obligatory set-aside to qualify for subsidy payments on the remaining land. Land diversion has been common in the US since the 1930s where its primary objectives have been to reduce soil erosion and groundwater depletion (Potter, 1991). In the EU, by contrast, the purpose of the present policy (and of a previous voluntary set-aside scheme in operation from 1988 to 1992) is primarily to reduce the surplus of agricultural production without further distorting the world markets for the products.

According to the EU policy, cereal producers who produce over 92 tonnes are only paid the guaranteed prices for their cereals if they designate 15 percent of their land to be set aside from production. In 1993 initial estimates suggest 600 000 ha of land set aside, making it the seventh largest 'crop' in the EU after wheat, barley, olives, fodder, vines and maize. The set-aside total compares with a maximum in 1991 of only 130 000 ha when under a previous scheme, set-aside and guaranteed prices were not linked and the scheme was essentially voluntary. However, the main objective of the present policy, i.e. to reduce surplus production, is not likely to be effective because, despite safeguards, the land set-aside tends to be of the lowest quality and yield potential (Buckwell, 1992). As a result the previous set-aside scheme on 2 percent of arable land in the UK in the five years up to 1992 brought about a reduction in production of cereals of only 0.75 percent.

Although the previous discussion in this chapter has suggested that 'surplus' land is a misnomer, the agricultural land set aside from production in the UK and in the EU is *theoretically* available for afforestation. What then would be the potential for augmenting greenhouse gas emission reduction through the use of this land 'surplus' to agricultural production? The cost of afforestation can be investigated to give a maximum cost per tonne of carbon sequestered. This is an upper-bound estimate because: (a) a complicated system of subsidy exists in agriculture, and there would be a resource saving in terms of reduced producer subsidies (indeed this is explained in Chapter 6 with regard to other agricultural policy options); and (b) the estimates do not account for the differential financial returns to forestry and agriculture.

If land has been set aside from agricultural production the value to society of the agricultural production previously taking place on this land is implicitly zero, or even negative. However, other costs associated with afforestation, such as land purchase costs, will be artificially high due to the general level of price support to agriculture. The following section investigates these distortions to the true social costs of this abatement option.

The costs of afforestation

Two stylised forestry options for afforestation of surplus agricultural land in the UK are possible: first an option which has the carbon sequestration properties of forestry as a primary objective; and secondly an option which takes account of amenity and other objectives of afforestation. If *maximising* carbon sequestration benefits of forestry were the primary consideration, then from the analysis set out above, a short-rotation species able to be utilised for bio-energy would be most desirable. In the conditions prevailing in the UK coppicing of poplars could provide such an option. Cannell and Cape (1991) estimate that 3 percent of UK emissions would be offset by 1 million ha of poplars.

However, given the infrastructure costs of biomass energy and the growing market for pulp and other wood products, it is more realistic to consider the option of conifer-based afforestation, along the lines of the predominant afforestation in the UK in the last half century (see Chapter 2). The afforestation scenario examined below is the simple case of 100 percent sitka spruce, assumed to have yield class 12 (which is 12 m³ ha⁻¹ year⁻¹ of merchantable timber over bark at the age of maximum mean annual increment).

Afforestation in the last half century in the UK has been criticised by many for promoting the single objective of commercial timber to the detriment of other objectives such as amenity and landscape enhancement (Mather, 1993). In recent years the planned new afforestation in the UK has largely been in urban fringe where these multiple objectives have been pursued (Bishop, 1992), especially after the restructuring of the tax system to reduce the implicit subsidy of private forestry investment prior to 1988 (Crabtree and Macmillan, 1989). A more acceptable option for afforestation of surplus agricultural land from the perspective of multiple benefits, would be of broadleaved woodland of mixed species. The broadleaved option investigated below is of afforestation with the same species mix as the existing broadleaved forest estate in the UK (see Chapter 2), namely a mix of oak, beech and sycamore, ash and birch (SAB).

The carbon sequestration rates and equilibrium carbon storage for these options (CON and BRD) are shown in Table 7.3. Coniferous forests accumulate carbon at a faster rate than broadleaved forest (3.7 compared to 1.28 t C ha⁻¹ year⁻¹) but in equilibrium store less carbon, given the assumed 12 m³ ha⁻¹

Table 7.3 Estimates of carbon sequestration for UK afforestation

	Conifer afforestation option (CON)	Broadleaved afforestation option (BRD)
Species mix	100% Sitka spruce YC 12*	Oak (39.7%), beech (17.1%), sycamore, ash and birch (43.2%)
Equilibrium carbon (t C ha^{-1})	68	140
Annual rate of storage (t C ha^{-1} year^{-1})	3.7	1.28

Sources: Based on Adger et al. (1992a); Dewar and Cannell (1992); Hamilton and Christie (1971).
*Assumes no thinning regime and 2.0 m initial spacing (see Dewar and Cannell, 1992).
Notes: Rate of storage is total C storage at end of first rotation divided by rotation length. Changes in soil carbon due to afforestation are not included in the estimates.

year^{-1} yield class average for the UK. The equilibrium carbon sequestration for these two options on the whole area of set-aside in the UK are, 2.22 mt C year^{-1} from CON and 0.77 mt C year^{-1} from BRD. These constitute 1.4 and 0.5 percent respectively of the UK's estimated carbon emissions from CO_2 only in 1990.

What would be the cost of these two afforestation options? The CON and BRD options have different establishment and maintenance costs and have different revenue profiles, as the trees mature at different rates. Table 7.4 shows the assumptions made to estimate the net present value of the two options. The calculations are essentially the same as those presented above for the US by Sedjo and Solomon (1989): establishment, maintenance and land purchase but taking the future revenue from timber into the equation. Costs are averaged for the UK (Nix, 1993) with land costs of vacant possession of Grade II arable land being taken as an average of the land set-aside from arable production. The net present values of CON and BRD are both negative, highlighting the point that forest subsidies play a large part in afforestation decisions in the UK because the opportunity costs of foregone agricultural production are high.

The costs of sequestration, the estimated *abatement* costs, are £21.47 per t C for CON and £58.08 per t C for BRD. The least-cost option is therefore CON, but as discussed above, broadleaved afforestation has higher amenity value, and indeed it is unlikely that large-scale coniferous afforestation in lowland Britain would be given planning permission. Despite being based on much lower growth rates than the work of Sedjo and Solomon (who assume 15 m^3 ha^{-1} year^{-1} for afforestation), the cost estimates presented here are lower than the Sedjo and Solomon (1989) estimates of approximately £85 per t C sequestered. The difference is due to the inclusion of *revenue* from timber production in this study. Current prices were assumed to be constant over time, an assumption also made by Pearce (1991b). It is further assumed that afforestation of 0.6 million ha in the UK will have a

Table 7.4   Costs and carbon sequestrations for afforestation options on UK set-aside land

	Coniferous afforestation option (CON)	Broadleaved afforestation option (BRD)
Species mix	100% Sitka spruce	Oak (39.7%), beech (17.1%), sycamore, ash and birch (43.2%)
Establishment cost	£1500 per ha	£2000 per ha
Land purchase	£3600 per ha	£3600 per ha
Present value of costs	£6549 per ha	£6517 per ha
Present value of revenue	£2577 per ha	£2800 per ha
Net present value	−£3972 per ha	−£3717 per ha
Total cost to afforest 0.6 million ha	£2383 m	£2230 m
Sequestration cost per tonne of carbon	£21.47 per t C	£58.08 per t C

Sources and notes: Establishment, land purchase and other costs (maintenance, fertiliser) and timber prices based on Nix (1993). Land purchase based on arable agricultural land (Grade II) bought at vacant possession (Nix, 1993).
Total costs are based on 1993 estimates of 0.6 million ha of set-aside land (see text).
Sequestration rates are based on Table 7.3. Net present values of afforestation based on 3 percent discount rate and 50 year time horizon, with coniferous felling at 35 years. NPVs are negative and are therefore only financially attractive with subsidy (see text).

negligible impact on prices, and hence the UK is a 'price taker' for timber. Sedjo and Solomon (1989) could not make this assumption as afforestation of 465 million ha would have a dramatic impact on future timber prices.

Social cost of afforestation

The estimated abatement costs above are the costs of the afforestation options given the predominating market prices. The NPVs are negative for both options which implies that no farmer would afforest the set-aside land without financial inducement. Indeed the present UK afforestation grants for arable land are up to a maximum present value of £7532 per ha for CON (Woodland Grant Scheme and Farm Woodland Premium Scheme) and £8517 per ha for BRD, given highest rates and multiple availability of all subsidies over a 50-year horizon (calculated from Nix, 1993). The net exchequer cost is the difference between afforestation grants and set-aside. Set-aside depends on the previous crop and can range from present values (with the same 50-year time horizon) of £5000 – £11,400 per ha (calculated from MAFF, 1993b). The extra cost to the exchequer of afforestation may be minimal. However, the agricultural price support system means that the subsidies for afforestation would be borne by the national exchequer and those of set-aside by the EU. The non-integration of policies therefore distorts the decision over subsidies from the government or exchequer viewpoint. We shall return to this point below.

To evaluate the true social costs of afforestation all the environmental impacts of afforestation on amenity, recreation, pollution and its opportunity costs would need to be considered. Two such studies—for the UK (Pearce, 1991b) and for Scotland (Macmillan, 1993)—exist, both of which include the *benefits of carbon sequestration* in terms of global warming damage avoided (see Chapter 6) in their full social cost–benefit analysis. The studies show a range of options for forest management which reach the desired social rate of return, but that for a significant proportion of Scotland's forests, natural regeneration is a more desirable option than replanting cleared forest (Macmillan, 1993). The multiple objectives of forests then could well be in conflict with a single objective such as carbon sequestration.

In this study we will not undertake a full social cost–benefit analysis. Rather we will demonstrate that the financial costs of carbon abatement through afforestation are upper bound estimates of the cost to society in the UK example. This is done through adjusting the market prices for the inputs and costs associated with forestry. Specifically, the market price of agricultural land is inflated due to the system of agricultural support, which affects all land use decisions, from the development of agricultural land to built environment, to recreation and forestry. Harvey (1990) estimates the price determinants of agricultural land in England and Wales using time series data. From previously calculated estimates of the proportion of agricultural produce made up of subsidy (PSEs; see Chapter 6), he calculates that the *shadow price* of land may be only 56 percent of the present market price. This result means that the real cost of using agricultural land for afforestation is significantly less (only 56 percent) than in the estimates made above. However, Harvey (1990) signals that simply applying this factor to land prices will *underestimate* its social cost, as the market price of land would drift downwards as agricultural support declined, till the market and social value converged. Further, the demand for conservation, amenity and rural environment goods is high so if set-aside land became available for non-agricultural purposes, this demand would ensure that its true cost would be higher than the 54 percent of market price which reflects the price if all agricultural price support were removed. Based on these arguments, Pearce (1991b) takes two estimates of the social cost of land for afforestation: 50 percent and 80 percent of the market value. Taking this range of the social value of land gives the range of abatement costs of carbon through afforestation shown in Table 7.5.

The influence of the assumptions about the real cost of land gives the relationship shown in Figure 7.1. In summary, the abatement costs of coniferous afforestation in the UK context are lower than those from broadleaved afforestation, no matter whether the financial, exchequer or 'real' costs of afforestation are taken. Assessment of afforestation options are rarely based on this single objective. Nor should they be.

Table 7.5   The cost of carbon sequestration through afforestation of set-aside land in the UK: comparing the financial with the social cost

	Afforestation option	
	CON	BRD
Financial cost land at market value	£21.47 per t C	£58.08 per t C
Social cost assuming land 50% of market value	£11.67 per t C	£29.75 per t C
assuming land 80% of market value	£17.55 per t C	£46.75 per t C

Will afforestation come about?

The previous section has estimated that a small proportion of the UK's greenhouse gas emissions could feasibly be offset through the afforestation of agricultural land. The cost of this to society could be as low as £11 per t C, though the cost to the government in subsidies would be likely to be high. This raises important issues regarding agricultural policy which have been at the forefront of the present CAP reform:

First, is the support of farming incomes a legitimate purpose of agricultural

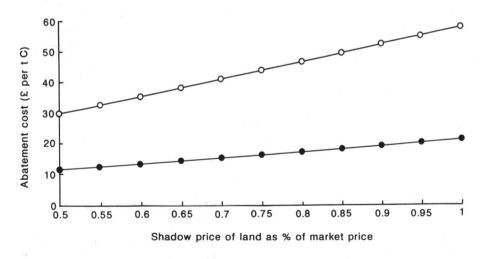

Figure   7.1   Relationship   of   social   cost   of   land   to   abatement   cost   of afforestation options in the UK

policy? Secondly, if it is legitimate (and it is enshrined in the Treaty of Rome which formed the European Economic Community in 1958), then what mechanisms should be used to ensure that rural landowners deliver the *environmental* benefits presently demanded?

In this context, offsetting the greenhouse effect would seem to be a legitimate objective in the use of non-built land. However, although the objective has been recognised in both UK and EU policy documents, the present agricultural policy reforms militate against afforestation for a number of reasons. These are primarily that owners already receiving subsidies for forestry cannot be counted as part of set-aside; that set-aside obligations are less if the area is rotated, precluding afforestation; and most importantly, existing afforestation subsidies are not sufficient to encourage large-scale transfers of land. The underlying reason is that the set-aside policy was introduced to *reduce agricultural production* not to fulfil other 'environmental' policy objectives. Specifically the present set-aside policy in Europe undermines afforestation for the following reasons:

— Farmers qualify for price support on arable produce only if they agree to set-aside 15 percent of their land which was under arable crops in 1991. A per hectare subsidy payment also accrues on the set-aside land. In order to ensure that the least productive land is not set-aside, the 15 percent set-aside must be rotated. This precludes afforestation. If the farmer chooses not to rotate the set-aside land, 18 percent of the arable total must be set-aside to attract the subsidy payments.
— Land which attracts subsidy under UK national afforestation schemes in general cannot be counted as part of the set-aside land. Further, land in a previous voluntary set-aside scheme running from 1988 to 1992 could not be counted unless it was taken out of set-aside. If it had been afforested it would need to be ploughed up and planted with arable crops to enable it to be counted as part of the total arable area.
— Set-aside land only has to be within the same yield area to be counted by a single farmer. In the UK case one yield area covers the whole of England. A market in renting land that was under arable crops in 1991 has developed to allow large farmers to qualify for subsidy payments while retaining economies of scale. The short term nature of rental agreements precludes afforestation on this land.

These specific reasons militate against the afforestation of set-aside land, but as mentioned above the underlying reason is the entrenched nature of agricultural support to the resulting detriment of other environmental objectives including greenhouse gas reduction. The CAP reform process had, as its primary objective, the reduction in the budgetary cost of support and the targeting of support to smaller farmers. The resulting cost saving was to be redistributed in rural areas through 'agri-environmental' measures. As the UK and France insisted that the subsidy for set-aside land should be paid on all land (rather than the proposals by the EU which would have limited this subsidy to the first 50 ha of set-aside), the cost of the policy escalated, thereby pushing back the priority given to agri-environment, forestry or other measures. Similar circumstances can be found when examining the EU's previous attempt to divert land from agriculture through set-aside payments from 1988 to 1992. Only a small proportion of this land was afforested.

7.5 CONCLUSIONS

Afforestation of large tracts of land which are assumed to be 'surplus'—either because they are not obstensibly used for agricultural production (as in the case of many areas in tropical countries) or because they are being taken out of production (in the EU and the US)—appears to be an attractive land-use option to sequester CO_2. Some studies reviewed here indicate that such strategies represent least-cost abatement options. However, these studies have assumed that land can be obtained at zero cost; in other words that it has no opportunity cost. We have demonstrated that this is not the case. Furthermore, the studies which have made estimates of the possible magnitude of afforestation for carbon sequestration on a global or cross-national scale, give very scant attention to the likely feasibility of such schemes. In particular, issues such as property rights are not considered in these analyses.

Discussion of the capacity and viability of using set-aside land in the UK for afforestation demonstrates that this would be extremely unlikely without subsidies and change in the current set-aside policies which appear to oppose afforestation as a land use option, partly because of its comparatively long time horizon. In conclusion, estimates of the technical feasibility of large-scale afforestation to offset the greenhouse effect are dependent on defining and identifying available land or land 'surplus' to present needs, then estimating the potential carbon sequestration of these. These results are meaningless in policy terms without a consideration of both the reason why forests have been lost in the first place and the mechanisms by which afforestation can take place. In Chapter 5 we explained that deforestation often takes place for reasons of economic necessity. The predominant subsequent use of forest land is for agricultural purposes but in many cases this is carried out because of distorted political or economic signals from government or global markets. So although removing these distortions could halt or substantially reduce deforestation, the concept of 'surplus' land available for afforestation is seriously discredited. Each land use must be examined in its political and economic context, and strategies of greenhouse gas emission reduction based on afforestation must consider the multiple benefits and costs of forests as a land use.

CHAPTER 8

The International Policy Dimension

8.1 INTRODUCTION

Given the preceding discussions on the causes of deforestation and possible domestic policy solutions, and the role of agriculture and land use policies in reducing other emission sources, this chapter examines further opportunities for the reduction of emissions. As has been shown in the previous chapters, it is the open access nature of the atmosphere which leads it to be used unsustainably: the decision-makers who cause emissions create environmental impacts for which they do not suffer the consequences. There therefore exists the opportunity to compensate these decision-makers for the opportunity cost of not emitting. This process necessarily takes place at the international level, and hence, the international impetus provided by the Climate Change Convention is crucial in identifying cost-effective transfer mechanisms to assist emission reduction and ultimately to avoid the impacts of global climatic change. Likewise the effectiveness of these mechanisms may well be critical in successful implementation of the Convention.

Policies to reduce greenhouse gas emissions by definition have an international dimension: both the impacts of global warming and the costs of emission reduction are spatially skewed across the world: different countries will experience different costs and benefits. The Climate Change Convention has set targets for emission reduction and suggests ways in which countries can co-operate in this, and in adaption to climate change. The content of the Convention has already been set out in Chapter 1, but its influence on emission reduction policy is further expanded upon here. The funding of resource transfers is the key to successful compliance with the Convention, mainly because the impacts of climate change are likely to fall on countries with the least resources to cope (Barbier and Pearce, 1990), but also, in the context of this study, because the opportunities for land use related emission reduction may be largest in tropical forested countries and other areas with low per capita income.

The resource transfer clauses in the Convention deal with funding both the *costs of adaptation* to climate change and the *costs of complying* with the Convention. The former accepts that some impacts of climate change are already committed to and that the signatories should act co-operatively to avoid potentially catastrophic consequences. These include impacts on food security and on the low-lying coastal regions of the world. The commitment does not require the transfer of funds to any large extent at present, as the costs of adaptation to climatic change have yet to be incurred, with the significant exception of the present costs of increased insurance and under-writing of development in some low-lying states and regions (Leggett, 1993). The commitments for funding compliance with the Convention are immediate, however, and it is these costs which have created the greatest controversy. This chapter highlights these issues and the relevance of the Convention, as signed, to other potential policy instruments.

Resource transfers could come about through the disbursement of revenue from an internationally administered carbon tax, through a system of tradeable permits, through international offset agreements or through an international fund where payments are made on some assessment of responsibility. A carbon tax would raise revenue from a levy on energy or on the carbon content of fuels and other emission sources, which would then be available for expenditure on further emission reduction technologies or projects. A tradeable permit system entails the creation of a market whereby each country is given an initial allocation right to emit, and if surplus emission rights exist, these could be sold to other countries which exceeded their quota, thereby facilitating the transfer of resources from profligate emitters to those countries which emitted little. International offsets are bilateral trades of emissions rights, whereby one country reduces its emissions at less cost than it could domestically by investing in emission reduction or sink enhancement in another country.

The advantage of these mechanisms is that they ensure that emissions are reduced at low resource cost. Cost-effectiveness is the major criteria advocated by economists in deciding on policy action on reducing emissions. However, equity and fairness are also important: national policy-makers and country representations to international accords require that policies are seen to be fair for them to be acceptable. The international aspect of this issue is manifest in the debate over responsibility for global warming. This chapter sets out the debate for two aspects of responsibility: *historic* responsibility for the presently observed atmospheric concentrations; and the differentiated *present* responsibility for emissions from those on low incomes and for land use related sources. The equity position adopted in operationalising solutions will fundamentally determine the transfer of funds in any international initiative to provide resources and incentives for emission reduction.

Theoretically, there are advantages to the market-based policy instruments, as emission reduction takes place at least resource cost. However, in practice only international offsets have even been seriously considered as feasible and this chapter will show that even with offsets there are serious drawbacks if these are undertaken in the land use sector, primarily due to the transaction costs involved in operating the projects, the potential environmental impacts of plantation forestry in the tropics, and also due to the issue of sovereignty and responsibility.

8.2 CLIMATE CHANGE CONVENTION

The Climate Change Convention commits its signatories to stabilising their global emissions of greenhouse gases at 1990 levels by 2000 so as to avoid dangerous anthropogenic interference with the climate. The necessary actions are to be undertaken by all countries, though not at the expense of poverty alleviation and necessary economic development, so long as this development is sustainable. The signatories were classified in the annexes to the Convention as: *industrialised countries*, which have to take all the actions and to agree to the funding clauses; *developing countries*, which have only to comply without jeopardising their economic development; and *economies in transition*, from the former centrally planned countries, which have an unspecified time allowance before having to implement the agreement.

The industrialised countries have agreed that they will fund the costs of compliance with the Convention of developing countries, but the extent of the funding depends on the particular costs. Developed country signatories have agreed:

(a) to pay *full agreed costs* for drawing up national inventories of greenhouse gas sources and sinks and other information to be conveyed to the Conference of the Parties (Article 12, Para. 1);

(b) to *assist in meeting the costs of adaptation* to the adverse effects of climate change (Article 4, Para. 4).

(c) Controversially, however, for the major abatement costs (promotion of technology transfer, sustainable management of sinks and reservoirs, public awareness and training related to climate change, coastal zone management; Article 4, Para. 1), developed country parties have agreed to fund the *full incremental costs* of compliance with the Convention. This is the area in which there is greatest scope for interpretation.

The Convention designated the Global Environment Facility (GEF) as the manager of the funding mechanism for the Convention, until the Conference of the Parties to the Convention meets and implements the funding mechanism itself. The GEF also has responsibilities under the Biodiversity Convention, as well as for funding projects to reduce emissions of ozone-depleting substances, and for the protection of international waters. A major criterion for funding initiatives under the Climate Change Convention is that they will reap benefits to the *global* environment (GEF, 1992c; Pearce and Warford, 1993). As such, the GEF funds projects which would not normally be undertaken as domestic costs exceed domestic benefits, and hence the global environmental benefits would not be realised. For example, if afforestation that yielded greater domestic benefit than cost was proposed, then the extra costs associated with the project would not be considered for funding by the GEF, even if greenhouse gas sink enhancement was achieved. The project would be funded, however, if the project could not bring a net return to the domestic economy and hence required additional funding—the incremental cost—for the global environmental benefits to be realised.

Incremental cost is an ambiguous term which can be interpreted in a number of different ways. On the one hand, it may be equivalent to the extra or marginal cost of compliance within the terms of the Convention, regardless of whether costs and benefits accrue nationally or globally. On the other hand, the GEF interpretation includes funds only up to the value of those benefits accruing at global level. This ambiguity, and others within the Convention, have been blamed on hasty re-writing of the text of the Convention in 1992 (Drennen, 1993).

Another of the issues open to interpretation is that of *joint implementation*. As discussed below, joint implementation can be carried out through bilateral agreement or through multilateral funding. These are essentially different in terms of where, or to whom, the benefits of the reduction accrue. It is not clear whether the Climate Change Convention can allow the enhancement of sinks made as a result of bilateral agreements to count as part of the donating countries' inventories. Whether this occurs or not will depend on the outcome of negotiations by the Conference of the Parties to the Convention.

The Climate Change Convention sets out the possibility for countries to co-operate with each other in compliance strategies:

> ... each of the parties shall communicate ... detailed information on its policies and measures ... with the aim of *returning individually or jointly* to their 1990 levels of these anthropogenic emissions of carbon dioxide and other greenhouse gases ... [our italic].
>
> (Article 4, Para. 2b of Climate Change Convention)

This could be interpreted as the Parties having to enter agreements so that the aggregate emissions from the two countries would be counted together. This is clarified in Article 12, Para. 8, which places the ultimate responsibility on the individual governments:

> Any group of parties may ... make a joint communication in fulfilment of their obligations ... provided that such a communication includes information on the fulfilment by each of these parties of its individual obligations under the Convention.
>
> (Article 12, Para. 8)

The Convention then allows the trading of emissions within economically integrated regions if this is desirable. It also seems to allow those countries giving bilateral assistance specifically for the enhancement of greenhouse sinks in developing countries to claim this as part of their net emission reduction strategy, though Article 12 above does reconfirm the importance of individual obligations, and hence joint implementation is not definitive.

8.3 RESPONSIBILITY FOR GLOBAL WARMING

The issue of responsibility for global warming is essentially political in nature, and fundamentally affected the negotiating positions of the parties to the Climate Change Convention. The climate belongs to everyone in that no-one can be excluded or can exclude others from its use. The major negative *impacts* of climate change, however, are likely to be significant only to specific people, and are not equally distributed. The impacts may well be greatest in nations with the least resources to mitigate them. Responsibility for *causing* the problem is a lever for allocating resources to alleviate the potential impacts, but also to instigate abatement measures. The Convention recognises that responsibility is a major issue:

> Noting that the largest share of historical and current global emissions of greenhouse gases has originated in developed countries, that per capita emissions in developing countries are still relatively low and that the share of global emissions originating in developing countries will grow to meet their social and development needs.
>
> (Preamble, Para. 3)

Under the Climate Change Convention as ratified, industrialised countries have agreed to undertake to fund adaptation to the adverse effects. Article 2, Para. 4 of the Convention sets out this responsibility thus:

> The developed country Parties ... shall assist the developing country Parties that are particularly vulnerable to the adverse effects of climate change in meeting costs of adaptation to those adverse effects.

The debate on responsibility for global warming focuses on a number of areas concerning:

(1) present emissions: uncertainty in natural sources and sinks, and whether all sources and sinks are relevant; and
(2) past emissions and historical responsibility.

Within the debate apparently scientific questions are being raised which underlie the political points. They focus on uncertainty in measuring the emissions of particular greenhouse gases and on the role of these gases in causing the greenhouse effect. The role of past industrial development in the cumulative atmospheric concentration of greenhouse gases highlights historical responsibility. A secondary related issue is whether certain emissions, i.e. those of low-income groups in developing countries, could be classified as 'survival' emissions rather than 'luxury' emissions. First, these emissions could not be easily reduced, and secondly, for equity considerations, they should not have to be reduced before the 'luxury' emissions. Greenhouse gases are ambient pollutants, so, for example, 1 t of CH_4 emitted from a waste disposal site in the US, has the same greenhouse effect as 1 t of CH_4 emitted from livestock kept by subsistence herders in Mali. Each emission makes an equal contribution to the greenhouse effect, but does each country have the same responsibility to reduce the emissions? We have already discussed uncertainty in emissions estimates and now consider the other areas of dispute.

Should all emissions be counted?

Where responsibility for global warming is compiled in an index, there is usually a problem in that all the gases are not accounted for—tropospheric ozone, carbon monoxide, halocarbons and nitrogen oxides (other than N_2O) are often ignored, either because of their low aggregate emissions or because their role as greenhouse gases is not well understood or because their inclusion may lead to double counting. This may distort the apparent responsibility for the overall effect.

A fundamental question is whether all sources of emissions of greenhouse gases should be taken into the calculus on emission reduction. If emissions from all agricultural or land use related sources were omitted, then policy measures to reduce emissions from these sources may be ignored, despite the fact that they may well be the most cost-effective strategies. In addition such policies may have other environmental and social benefits. Since many of the cost-effective mitigation options may possibly be found in developing countries, assistance by industrialised countries to undertake these options may prove beneficial to both parties. Further, the Climate Change Convention states that *all* sources and sinks are important and will be brought into the reporting process: all signatory parties to the Convention shall:

> Develop, periodically update, publish and make available ... national inventories of anthropogenic emissions by sources and removals by sinks of all greenhouse gases not controlled by the Montreal Protocol.
> (UN Framework Convention on Climate Change, Article 4, Para. 1a)

Historical responsibility

The Climate Change Convention has focused on present emissions but the issue of historical responsibility is still to be resolved if resource transfer on some equitable basis is to take place. The arguments put forward for concentrating on present emissions are that these are measurable, and that policies have to be devised to reduce them; secondly, that historical emissions of greenhouse gases took place out of ignorance of the consequences, and this accounts for the profligate use of fossil fuels and deforestation in industrialised regions; and thirdly, that the scale of activity causing the greenhouse effect was much less during the industrial revolution, and now that there is a greater burden on the atmosphere, each nation is responsible. Although these arguments do not make a valid case for ignoring historic responsibility, the fact that the developed countries have in theory used up a greater proportion of the atmospheric sink, whether ignorant of the consequences or not, is often deployed as a rationale for resource and technology transfer to industrialising countries.

Indices of responsibility

Attempts have been made in recent years to account for some of these factors by producing indices of responsibility for global warming, although these have been controversial and often criticised as politically motivated (e.g. World Resources Institute, 1990; Agarwal and Narain, 1991; Ahuja, 1992). Recent estimates of emissions by selected countries of some greenhouse gases are presented in Table 8.1 along with the ranking of countries for highest total emissions of all gases based on estimates by the World Resources Institute in 1990 (WRI, 1990). In this, the United States has largest emissions and is ranked 1. This shows the range of emissions and the difficulty of reducing the information to a single allocation of responsibility.

Table 8.1 Greenhouse gas emissions of selected high and low income countries and ranking by total emissions

	CO_2* (mt C)	CH_4 (thousand tonnes)	N_2O (thousand tonnes)	CFC[†] (thousand tonnes)	WRI Index
Australia	61	5	20	16	19
Bangladesh	4	8	6	1	41
Brazil	270	126	133	14	3
Canada	32	3	37	22	12
China	494	52	228	16	4
Côte d'Ivoire	22	0	5	1	23
Ecuador	29	1	18	1	42
Egypt	23	1	6	6	46
Germany	270	6	72	80	7
Indonesia	158	9	41	17	9
Mexico	134	4	36	11	13
Netherlands	49	1	10	10	24
Poland	122	6	32	19	15
Qatar	4	112	1	0	—
Saudi Arabia	53	1	11	1	25
United Kingdom	154	5	44	56	8
United States	1246	33	756	375	1
Venezuela	37	1	12	5	35
World	6432	352	3783	1369	

*CO_2 in million tonnes Carbon equivalent.
[†]CFC in kilotonnes CFC 11 equivalent.
Sources: Emissions data from Subak et al. (1993). WRI Index shows the rank of the country in total net emissions as discussed in the text (World Resources Institute, 1990).

The ranking of countries by emissions using the WRI methodology has been criticised on the grounds that the assumptions made and data sources used always increased the allocation of responsibility on developing countries. A central dispute is over how the different gases were weighted. WRI, however, claimed that their methodology avoided this issue by comparing gases on an instantaneous rather than on a historical basis. This is in effect an arbitrary time horizon, since the GWPs of the gases are different for different time horizons, a point made by the Intergovernmental Panel on Climate Change 1990 report. A long time horizon underemphasises emissions of gases with short atmospheric residence times, whereas a short time horizon fails to take account of the future importance of long-lived gases such as CFCs. The WRI use of instantaneous effects therefore discriminates against those countries (typically non-industrialised) which are not major emitters of CFCs.

The difference in profile of greenhouse gas emissions is graphically represented by the case of New Zealand, which although not in the top 50 emitters in the WRI estimates of total net emissions, has the 15th highest per capita output of total greenhouse gases (using the WRI methodology). The reason for this is the large emissions of methane from ruminants. Of New Zealand's aggregate contribution to the greenhouse effect, 61 percent comes from methane emissions (based on estimates of Subak et al. (1993) and comparing the gases on a direct global warming potential with a 20-year time integration horizon), making it the third highest per capita emitter of methane in the world. Similarly Qatar, shown in Table 8.1 to have low aggregate emissions, and not ranked in the highest 50 emitting nations, is in fact the second highest in per capita emissions.

The WRI estimates are of net emissions calculated by taking annual emissions over the airborne fraction (change in atmospheric concentration/ total emissions in time period). This has been discredited both for being unstable and providing counter-intuitive responses (it is possible for a country's greenhouse index to go up when its emissions are falling); and for implicitly allocating the total global sink for emissions proportionately to gross emissions.

It is claimed that the WRI used data on deforestation for 1987 which were unusually high for Brazil and as a result 'promoted' Brazil to being third in the greenhouse index behind only the USSR and the United States. Further, using only present emissions implicitly assumes that countries hold no historical responsibility, which is implicit in the Climate Change Convention. As for future emissions, the case has been made that the projections of responsibility for *reducing* emissions also unfairly discriminate against developing countries. The climate change scenarios of a 20 percent reduction of greenhouse gas emissions, by IPCC (Houghton et al., 1990) for example, are based on 'business as usual' reference scenarios which assume only industrialised countries will increase their emissions significantly in the future, but recommend that the burden of emission reduction should be shared by all (Parikh, 1992).

Alternative methods of allocating responsibility to those outlined above can be designed. Emissions only could be counted and emission quotas allocated on a per capita basis to the world's population, bringing about resource transfers to low-emitting densely populated countries. The debate about the allocation of responsibility for global warming is essentially political, so questions as to the allocation of quotas to bring about resource transfer to mitigate or ameliorate the effects of global warming on a global basis will only come about by international accord. Some of the other issues of dispute, such as the role of particular gases in causing the greenhouse effect, and emissions from natural sources, will be better informed with evolving scientific knowledge.

These issues will have, as the Convention is implemented, significant implications for some of the policy instruments, e.g. tradeable permits or international offsets, which are now described.

8.4 INTERNATIONAL POLICY INSTRUMENTS

The Climate Change Convention sets the same target for each country in terms of its emissions: to stabilise at 1990 levels by 2000. Whatever proposals were agreed to, an implicit allocation of responsibility for future action is involved as outlined above. Less developed countries see the present arrangement as inequitable, in that countries with high income have become rich through past pollution and emission of greenhouse gases. Non-industrialised countries therefore argue that their targets should be lower, and that their economic development is likely to be at the cost of increased emissions of greenhouse gases. To attract widespread support the Convention has therefore to be seen to be fair in terms of resource and technology transfer.

The present role of the GEF as the interim manager of the financial mechanism for greenhouse gas abatement is primarily to fund projects which would not otherwise be funded because they are too costly in the domestic setting, as discussed above. The GEF promotes international offset arrangements to reduce emissions and these will be described below. The GEF, however, is constrained in making significant impact on global emissions, especially in countries where resources and technology for emission reduction do not exist. Funds and technologies to reduce emissions require the parties to the Convention to institute some mechanism for transferring resources, such as an international carbon tax, a tradeable permit system or some other approach.

To attract widespread participation any mechanism would not only have to be effective, but be seen as fair. This raises a number of issues. For example, if taxation on carbon was proposed, would it be uniform across all countries? And what criteria would be used to compensate those countries with lower levels of economic development? Such questions have been instrumental in stalling the implementation of a carbon/energy tax in the EU. If an international tradeable permit system was introduced, what criteria would be used to set the initial allocation of permits? Various criteria have been suggested, such as per capita allowances; allocation on land area of the country; historic or current responsibility based on emissions; or ability to pay (e.g. Fujii, 1990; Grubb et al., 1992; Hayes and Smith, 1993b). We will now examine these policy instruments—carbon tax, tradeable permits, and international offsets—and some of these issues, in turn.

International carbon taxes

A tax on greenhouse gas emissions through taxing fossil fuel has many advantages. It would be cheap and simple to administer as taxes on hydrocarbons already exist. In theory a tax need not increase the overall tax burden as it could offset other taxes, such as those on income, and hence be fiscally neutral. It has been argued that a tax on carbon should be used to an even greater extent to reduce other taxes which provide perverse incentives, and thus a double dividend might accrue from a carbon tax (Nordhaus, 1993). It has also been argued that a carbon tax would have no long term effects on economic growth (see Barker, 1993 for a review).

Although a carbon tax may be an attractive mechanism for individual countries faced with the emission target of the Climate Change Convention, there are problems with making a carbon tax work at any international level. An international tax by definition requires the co-operation of individual countries. First, there will always exist the incentives for free riding, i.e. for a country to gain comparative advantage by not administering the tax while other countries do. Secondly, as has been shown with negotiations over a proposed tax in the EC, each country is unlikely to agree to the tax level being equal across countries with different levels of economic development and technology. The impact of a tax on total emissions may also be offset since with decreased demand for oil with the higher price, countries which export oil may lower the price before tax to maintain their levels of income and hence increase demand in countries outside the carbon tax system. If this happened, a greater tax level would be required to produce the same emission reduction (Barrett, 1991).

For an international carbon tax to yield revenue for disbursement to countries with presently low levels of economic activity, and hence to secure their agreement, it would require an international mechanism for setting the tax level and disbursing some of the pre-arranged surplus revenue. However, as noted above, the political acceptability of carbon or energy taxes would be greatly enhanced by the redirection of the revenue to offsetting other taxes. Further, the hypothecation of carbon tax revenue for emission reduction purposes is against the general principles of taxation in many countries. For these reasons, the funding of international emission reduction through the revenue from a carbon tax is a remote possibility.

Carbon taxes are simple to administer as long they are levied on production of the sources of emissions. However, coverage of all the processes (or products, if the tax is alternatively on consumption) involved in greenhouse gas emissions is impossible, especially with regard to non-point land use based emissions and fluxes. Indeed a subsidy, in effect a negative tax, would be required for significant sequestration activities such as forestry (Anderson, 1991), and across all gases based on their radiative forcing or other weighting system (see Chapter 1), if the strategy were to be comprehensive. One of the advantages of a tradeable quota system, as opposed to an international taxation solution, is in the predictability of the total emission loading; it is up to each individual country to determine its strategy relative to its sinks, land use and energy emissions.

Tradeable permits

A tradeable permit system would require the setting of a total acceptable emissions level and the allocation of the right to emit this level between all the countries in the world. In effect, the Convention does indicate an overall acceptable level of emissions: stabilised 1990 emissions. However, the allocation of rights would not necessarily be on the basis implied in the Climate Change Convention, which is essentially maintaining emissions at the same level for each country. An allocation could alternatively be made on per capita emissions, say allowing all people in the world (or all adults, to avoid perverse population incentives; Grubb, 1989) to emit 1 t of carbon equivalent per year. This would mean that populous countries would have surplus emission rights and could sell these on an international market.

A tradeable emissions system has many attractions, some of which are shared with the other main market based instrument discussed above, the carbon tax. A tradeable emissions system would encourage more technical progress in emissions control than would the existing target system. This is due to the countries involved having an incentive to develop and apply technologies even when they have complied with the nominal targets (stabilisation of emissions), since further abatement of emissions still has the value of the saleable entitlements and would therefore generate income.

The benefits of permit trading are shown for the simplified two-country example in Figure 8.1 (following Barrett, 1992). The total length of the horizontal axis measures the total emissions agreed (based on independently agreed criteria) as the acceptable level of emissions (for example decided by the Climate Change Convention). If initially country i emits $0.E^o_i$ and j emits (measured from the right axis) $0'.E^o_j$, then the total emissions exceed that level of emissions agreed upon. One or other of the countries must then reduce their emissions and this costs the country at the rate given by their marginal cost curves, MC_i and MC_j. As outlined above, for any policy instrument, allocation of rights, responsibilities and revenues is either an explicit or implicit decision. In the case of a tradeable permit system, the allocation is explicit and could be made on the basis of population; 'grandfathered' historic emissions; or other criteria. If, for example, the allocation was on a per unit of adult population basis, and meant an

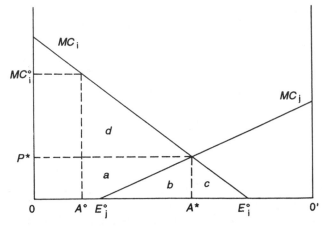

Cost to country	Before trading	After trading	Net effect (saving)
i	$a+b+c+d$	$a+b+c$	$-d$
j	0	$b-(a+b)$	$-a$

where $+ve$ = cost; $-ve$ = benefit. Based on Barrett (1992)

Figure 8.1   The gains from trading in a tradeable permit system with total allowable allocations of $0.0'$ and initial allocation A^o. The equilibrium allocation of emissions is at A^* and the price of permits is P^*. The gains from trade are shown.

allocation at A^o in Figure 8.1, then country j, having low per capita emissions, would not have to reduce its emissions at all to comply with the policy and *all* the emission reduction would have to be carried out by country i, a country with relatively high emissions per head of population.

The total cost of abatement to country i is given by the area under the marginal cost curve. So with allocation A^o, the cost to country i of reducing its emissions is given by $a+b+c+d$. Since the marginal abatement cost of country i at this point is much higher than that of country j, there is mutual benefit to trading; country i could pay country j to reduce emissions so that country i could avoid having to do so at higher unit cost. These benefits exist up to the point where the marginal cost for each country is the same (at A^*). The benefit to each country is shown under Figure 8.1. The benefit to country i is the area d, the difference between the initial cost of abatement and the sum of the abatement cost after trading and the revenue paid to country j. Country j receives revenue $a+b$ but incurs abatement cost b, so benefits by the area a. Hence given a system designed such that the initial allocation is acceptable and attracts all countries to participate in the agreement, trading will bring about mutual benefit and is desirable in a global sense since total costs would be minimised (Barrett, 1992).

Hypothetical estimates of the size of the transfers involved if the allocation

were indeed made on a per capita basis are given by Bertram (1992), illustrated in
Table 8.2. Bertram estimates that a large proportion of the world's emissions
rights, possibly over half, would need to be traded if the initial allocation was
made on a per capita basis. If 4.9 billion t C are emitted by 4.9 billion people in the
world, and this level of emission should be stabilised, then an allocation would be
1 t C per person. The results are informative, even if the empirical data and
assumptions can be criticised. The estimates of both emissions and population
are low for 1990; the allocation on a per head basis should try to avoid perverse
population growth incentives (Grubb, 1989); and the desirable overall emission
level could be set at some level other than stabilisation of present emissions. The
results show that the industrialised countries would need to purchase large
numbers of emissions permits, primarily from the most populous developing
countries such as China and India. On the estimates presented in Table 8.2, 48
percent of the initial allocation would need to be traded, bringing significant
revenue to the recipients. The price of a permit is determined by the demand for
the services brought about by emitting energy. Hence the price will rise or fall
depending on the availability of non-fossil fuel alternatives, and in the land use
context, on the relative profitability over time of sink enhancement compared to
the alternative uses of the land. Bertram (1992) hypothesises that with a permit
price of US$20 per year, equivalent to some already enforced carbon taxes, the
transfer would be in the order of US$50 billion.

The data presented in Table 8.2 may lead to the conclusion that the
industrialised countries would be unlikely to join an agreement which required
large expenditures on emission permits. A fundamental requirement for an
effective international agreement is one where all or many of the parties agree to
enter on the conditions laid down. A feasible system then would have to take on
board ability and willingness to pay. One other issue which affects both the
likelihood of agreement to a system and to its effectiveness in preventing
emissions, is the issue of what the revenue is spent on. If the only transfers are
revenue, then there will be an incentive in the developing countries to invest in
industrialisation, which, if the whole agreement is not strictly enforced, could
lead to greater emissions. If the development path chosen is one of industrialisation
based on new imported technologies, opportunities will exist for super-normal
profits in the medium term for the countries with these technologies, and hence
make the enforcement of the system beneficial to the high-technology countries.
Whatever the outcome, there is no guarantee that financing will result in a
desirable solution for all, and certain political and ethical questions remain.

International offsets

Although various tradeable pollution permit systems have been introduced for
emissions to river catchments and for air pollutants such as SO_2, primarily in the
US (see Hahn, 1989 for a review), they are at present a novel concept on the

Table 8.2 Hypothetical global permit-trading outcome for selected countries, based on per capita allocation

Country	Per capita carbon emissions (1980s) (tonnes)	Permit allocation, based on population (million tonnes)	Total carbon emissions (million tonnes)	Net buy-in required (million tonnes)
USA	4.9	241.6	1183.8	942.2
Canada	4.4	25.6	112.6	87.0
Czechoslovakia	4.1	15.5	63.6	48.1
Australia	3.9	16.0	62.4	46.4
Germany	3.3	77.5	257.9	180.4
former USSR	3.3	281.1	927.6	646.5
Poland	3.0	37.5	112.5	75.0
UK	2.5	56.7	141.8	85.1
France	2.0	55.4	110.8	55.4
Japan	1.9	121.5	230.9	109.4
Italy	1.5	57.2	85.8	28.6
Spain	1.4	38.7	54.2	15.5
China	0.5	1054.0	527.0	−527.0
Brazil	0.3	138.4	41.5	−96.9
India	0.1	781.4	78.1	−703.3
Other	0.5	1891.3	898.9	−1024.4
World total	1.0	488.94	4889.4	0.0
Net purchases by deficient countries				2351.6
Total net sales by holders of surplus permits				2351.6

Source: Adapted from Bertram (1992).
Notes: Emission and population estimates from late 1980s do not include emissions from deforestation. Brazil, where deforestation emissions peaked in 1987, would be in a less preferential position were deforestation included. Germany refers to amalgamated emissions and population of the two parts of pre-unification Germany. Negative figures in last column signify surplus of emission permits. Czechoslovakia is no longer a unified state.

international scene. However, the first step along this road, the trading of emissions rights, has already begun. Many of the initiatives have been related to forestry projects. Offset deals may be undertaken by private companies, or by governments as part of bilateral agreements or multilateral arrangements. International carbon offsets may be cost-effective in terms of reduction of carbon emissions achieved, and may also be one way to mobilise private capital to fund forest conservation. In theory, forest conservation, as opposed to afforestation, may bring about many other benefits. However, such international contracts are unlikely to be feasible or make a major contribution to the control of greenhouse gases. The reasons for this are monitoring, enforcement and scientific uncertainties, and the implicit change in property rights involved in 'selling' carbon sequestration rights. The present situation and the prospects for this to develop into a global market in emissions rights is discussed below with reference to interational offsets and land use emissions.

8.5 INTERNATIONAL OFFSETS AND LAND USE EMISSIONS

International carbon offsets involve the reduction of net greenhouse gas emissions by one agent somewhere outside its own national boundaries and jurisdiction. These offsets could be implemented by private companies, by national governments or brokered through multilateral agencies. They may be carried out by providing funding for emission reduction through technologies which offset emissions, or through enhancing sinks by protecting existing forests or through providing funding for afforestation programmes. Each of these has a different effectiveness in terms of greenhouse gas emissions offset, and also in terms of other environmental impacts and probability of success.

The parties involved

Private companies are presently involved in afforestation projects and in tropical forest conservation but the reasons for their involvement are diverse: for commercial investment, for publicity, or in order to conserve genetic resources. Private companies may wish to engage in international offsets for various reasons; either direct profit or market share interests would determine their involvement if we assume that firms act as profit maximisers. *Industrialised country governments* may wish to engage in bilateral arrangements in order to offset their greenhouse gas emissions. Overseas development policy could be further justified in terms of carbon offsets if it reduced emissions through forest conservation or other offsetting activities, but could also form part of a national greenhouse gas strategy under the Climate Change Convention if this is acceptable internationally. Norway, for example, plans to spend part of the revenue from a carbon tax on international offsets (see Table 8.5). This is through co-financing with the GEF projects which reduce emissions in Poland and in Mexico. The Mexican project provides energy-efficient lighting, compact fluorescent lamps (CFLs), and involves a grant of US$10 million and loans of US$10 million from the World Bank and from the GEF, so Norway's additional contribution of US$3 million 'buys' Norway the claim on a proportion of the emissions offset.

Private companies may wish to improve their public profiles as well as their profit margins by exhibiting environmental benevolence, to be perceived as what has been termed 'global good citizens' (Global Environment Facility, 1993). In this context, conservation of tropical forest projects are attractive because of heightened public interest in the industrialised countries in tropical deforestation (see also Chapter 3). Companies may also be anticipating national legislation on emissions which will allow offsets. This seems to be a major justification for investigating possibilities of offsets, e.g. by electricity generation companies in the USA. If state or national legislation gives credit for offsets, then these could already be in place. Further, it seems likely that national legislation will come about as a result of the Climate Change Convention, and as industrialised countries attempt to meet their commitments to the reduction of greenhouse gas emissions.

A few privately funded examples of forestry offset projects exist to date. An electricity utility in the US, for example, provided funds to an agroforestry project in Guatemala, through an aid organisation (CARE via USAID), specifically to offset its CO_2 emissions. The net effect of the project was calculated to offset the emissions of a proposed new power plant in the US, as estimated by Trexler et al. (1989). The project was to offset the emissions of 390 000 t C from a new 180 MW coal-fired power station. The AES calculations are conservative as to power station efficiency, making their emissions estimate relatively high as Freedman et al. (1992) estimate 343 000 t C emitted by a 200 MW coal-fired power station in New Brunswick. The forestry project chosen in Guatemala had begun in 1979 under funding from CARE, but was due to end in 1989 as a result of the termination of grant aid from USAID. Applied Energy Systems (AES), a US-based corporation, provided baseline funding of US$2 million as an endowment fund to finance training and operational costs, attracting funding from other agencies (US$13.1 million) and the Guatemala Government (US$1.2 million) totalling US$16 million. The project was chosen, according to Trexler et al. (1989) because it featured:

(a) a relatively high chance of successful implementation as organisational structures were already in place on the ground;
(b) the involvement of the local population who would benefit;
(c) opportunities for leverage of other funding.

Table 8.3 shows how the projected 16.3 mt C will be sequestered under the project over a 40-year period. The agroforestry scheme provides the bulk of the carbon accumulation and, importantly for local participation in the scheme, all of the activities enhance the value of land, either through safeguarding productivity in the case of soil conservation, or from income-generating opportunities in agroforestry.

An example of a proposed forestry offset project in Mexico is given in Table 8.4. The proposal was formulated by a union of *ejidos* (UCEFO) in response to a request from Applied Energy Services who already fund an agroforestry project in Guatemala to partially offset their carbon emissions. Ejidos are the Mexican organisations of co-operative land management and as such meet the criterion above of involvement and disbursement of benefits to local populations. UCEFO represents the group of ejidos that proposed the project, which is based primarily on a management scheme linking agriculture and forestry activities. The project is

Table 8.3 Carbon sequestration in AES Guatemala offset deal

Enhancement/sequestration activity	Details	Amount sequestered mt C
Soil conservation	3% accumulation in organic matter across eroded area of 8600 ha	0.3
Fire protection	2400 ha of additional forest protected and assuming 60% of area lost to any fire	0.6
Agroforestry	26.5 million trees planted on 66 000 ha of land	11.2
Woodlots	24.4 million trees planted to provide wood, adding to total biomass and displacing existing forest exploitation	4.2
Total		16.3

Source: Based on Trexler et al. (1989).

Table 8.4 Details of proposed carbon offset deal in Oaxaca, Mexico

Organisation	UCEFO (Union de Communidades y Ejidos Forestales del Estado de Oaxaca)
Area	114 600 ha total, approximately 10 000 ha forested with mixed pine and oak
Funding	$8.2 million total comprised of UCEFO $4.1 m Mexican Government $1.2 m Potenial donor $2.85 m
Proposed activities	Management plan linking agriculture and forestry activities: (1) Changing silvicultural techniques (Metodo de Desarrollo Silicola) to promote tree growth rates from 1.3 to 4.5 m^3 MAI by 2030 and increasing quality products by market research in pine resin products and further processing techniques. (2) Agricultural extension activities (3) Fire protection services
Net carbon sequestration	2.7 mt C by 2030
Cost of sequestration to donor	2.85/2.7 = >$1 per t C
Net cost of sequestration	8.1/2.7 = $3 per t C

Source: Based on Faeth et al. (1994).

similar to the Guatemalan agroforestry project in that the cost to the donor is low if the carbon sequestration of the whole project is claimed as the marginal abatement from the extra financing provided. Thus for the donor, if 2.7 mt C are sequestered by the project and the cost to the donor is less than US$3 million, then the average abatement cost is US$1–2 per t C. Overall the cost may be double that to the donor (US$3 per t C), but this is still a low average cost when compared to the options facing the donors, who tend to be electricity utilities in industrialised countries.

Table 8.5 shows this and other examples of greenhouse gas reduction projects already approved or implemented. The funding is from both private enterprises and national governments or through international agencies. The GEF project in Ecuador, funded under the third tranche of the GEF's pilot phase, is not a direct offset for a counterpart emitter. It is included in Table 8.5 as an example of multilateral brokering for reduction of *global* emissions.

As has been illustrated, private sector deals for carbon sequestration projects exist in several countries, both in the tropics and in temperate regions. Private companies may wish to offset their emissions in anticipation of emission regulations in their own countries or states or for the positive publicity involved in tropical forest protection. Examples of privately funded offset deals, and some other deals funded by state-owned utilities, are given in Table 8.5. These consist predominantly of a mixture of reforestation and agroforestry projects, which provide local fuelwood and income sources as well as displacing deforestation activities. The cost of carbon sequestration from new planting as well as from estimated reduction of forest losses varies in the examples given from < US$1.5 to US$30 per t C. Swisher (1991) analyses various projects in Central America and finds a range of US$3 to US$26 per t C (1989 prices).

Although forestry projects to offset greenhouse gas emissions through afforestation or conservation are considered here, it should be noted that this is only one rationale for forest conservation by private companies. Companies may have an incentive to conserve the biological diversity contained within tropical forests (Reid et al., 1993). Pharmaceutical companies have already undertaken forest conservation for the maintenance of genetic raw materials for potential products. For example, the US pharmaceutical company, Merck, has signed a contract with INBio, a non-government organisation in Costa Rica, whereby INBio supplies Merck with 10 000 samples of genetic material for screening, and INBio gains royalties on any commercial products developed from the supplied material as well as US$1 million over two years. Ten percent of this initial payment and 50 percent of the royalties in any products developed from the samples go directly to the Costa Rican Ministry of Natural Resources (Blum, 1993). Although no direct estimation of area of forest conserved through this deal is possible, carbon sequestration will occur as a secondary unaccounted benefit.

Table 8.5 Private and publicly financed carbon offset or enhancement deals

Company/organisation	Project	Other participants	Carbon sequestered (mt C)	Total cost ($ million)	Average cost ($ per t C sequestered)
Private forestry offsets					
Applied Energy Services	Agroforestry Guatemala	CARE via USAID, Govt. of Guatemala	15–58 over 40 years	$16 m	$9
Applied Energy Services	Agroforestry Paraguay	US Nature Conservancy	13 over 30 years	$2 m	< $1.5
New England Electricity Systems	Forestry, Malaysia	Rainforest Alliance, COPEC	0.3 –0.6	$0.45 m	< $2
SEP (Netherlands)	Reforestation, Malaysia	Innoprise	na (25 year project)	$1.3 m	na
Tenaska	Reforestation, Russia	Min. of Ecology, Russian Forest Service	0.5 over 25 years	~$0.5 m	$1 –$2
PacifiCorp	Forestry, Oregon US	—	0.06 per year	$ 0.1 m per year	$5
PacifiCorp	Urban trees, Utah	TreeUtah	na	$0.1 m per year	$15 –$30
Multinational sources					
GEF	Reforestation of degraded land, Ecuador	—	35	$2 m	$5.6
Non-forestry offsets					
Norwegian government	Mexico: energy efficiency Poland: coal to gas electricity conversion	World Bank	na	$4.5 m	na

Source: Dixon et al. (1993b), Trexler et al. (1989), Global Environment Facility (1992a, b).
Note: na is not available.

Carbon benefits of offsets

From an economic perspective, international offsets will be an attractive option if the cost of abating greenhouse gas emissions is lower through afforestation or through preventing deforestation, than through other strategies in the domestic country. The extent of the transfers involved is illustrated in Figure 8.2.

If the extra cost per tonne of CO_2 emission reduction for an industrialised country (i) is greater than that in a tropical country (t) over the range of emission abatement, then it is rational for i to abate in t up to the point where the extra cost becomes higher in i. Given the marginal cost curves of countries i and t in Figure 8.2, the industrialised country would benefit from reducing emissions in country t at lower cost per unit emissions up to emission reduction level e^*. The benefit gained is in the resources saved, which could therefore be directed into other uses, and which are shown by the area between the two curves (a). After the level e^* there are still benefits to trade but the transfer would be in the direction from t to i.

What evidence is there that the greenhouse gas emission abatement curves are the relative shapes depicted in Figure 8.2? Although all countries have the opportunity to reduce greenhouse gas emissions at negative marginal cost (Lovins and Lovins, 1992), the situation is reached where these costs turn positive for all countries and it is this which is portrayed in Figure 8.2. Greenhouse gas emission reduction through forest conservation and afforestation is ultimately a limited option. The limits are fundamentally the availability of land, given its

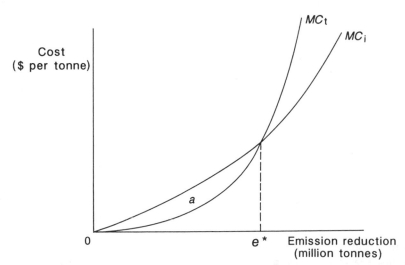

Figure 8.2   The benefits of international offset of emissions between industrialised country i and tropical country t. Lower marginal costs (*MC*) of emission reduction through forest conservation and afforestation in t provide incentives for i to reduce emissions there up to e^*. Revenue to country t of area *a* would result. Reproduced from Brown and Adger (1994) by permission of Elsevier Science

opportunity costs for other purposes. The marginal cost of the tropical country (MC_t) will then rise steeply when the low cost feasible projects are exhausted. Tropical afforestation has higher potential growth rates and therefore rates of carbon sequestration, and cheaper labour and land costs, and conservation of existing tropical forest can often be carried out at zero or low marginal cost if the distorting subsidies causing deforestation are eliminated. Dixon et al. (1993a) estimate average costs of sequestration through afforestation with a median range US$1–8 per t C in mainly tropical countries. However, these benefits may be to an extent compensated by investment uncertainty and risk in developing countries (Bekkering, 1992).

The gains will accrue to the industrialised country from having abated emissions at least cost and to the tropical country in the investment in forest resources; and the industrialised country will have the incentive to provide those resources because the emissions offset will be counted as part of its abatement effort. These benefits can therefore be appropriated *if* the Climate Change Convention is jointly implemented.

The distribution of the benefits (the area between the curves in Figure 8.2) will depend on the regulation of the transactions involved. Most of the benefit in terms of least-cost abatement may accrue to country i, but if the costs of abatement in tropical countries are very low, there would not be incentives for domestic abatement. An international regulator could set a higher market 'price' so that more resources than the actual cost of abatement would be paid by the industrial country and the extra resources could accrue to the developing country or be disbursed through the regulatory agency.

Other benefits of forest conservation

The benefits of the conservation of tropical forests, in terms of biodiversity, ecological functions of forests, and in terms of income generation for local people, also need to be accounted for in a comprehensive analysis. The concept of total economic value has been employed to quantify, at least from the human perspective, the benefits of forest conservation, of which carbon storage is one functional value component (Pearce, 1991a). The total *economic* value of a resource is made up of the price of the marketed products from the resource and from the non-marketed value, which is the services and goods from the forest or other resource that have no price or a low price due to their 'public good' nature (see Table 5.3). For forests, these non-marketed values include watershed protection and micro-climate regulation, a store for genetic diversity, amenity and landscape value as well as the carbon sequestration and storage.

The non-marketed environmental values (a subset of the total value) of Mexico's forests have been quantified (Adger et al., 1994). The distribution of these lower bound estimates between the components is shown in Table 8.6. This shows that carbon sequestration is a significant part of the non-marketed benefit,

Table 8.6 Estimated aggregate environmental values for Mexico's forests

	Use values			Non-use values		
	NTFPs	Tourism	Carbon	Watershed protection	Option	Existence
US$ ha^{-1} year^{-1}*	c. 330	—	20—103	—	6.4	—
Total m$ year^{-1}	6105.0†	32.3	3788.3	2.3	331.7	60.2

Source: Adapted from Adger et al. (1994).
*For a range of different forest types: tropical evergreen, tropical deciduous, temperate coniferous, temperate deciduous.
†Extrapolated from representative values for tropical evergreen and temperate deciduous forests only.

which includes the welfare gained by tourists, watershed protection services of Mexico's forests and the option value of future potential pharmaceutical discoveries. International offsets, or any forest conservation programme, produces these other joint benefits (calibrated here for the Mexican forest estate), rather than simply the carbon storage.

One important aspect in distinguishing the range of benefits from tropical forest conservation is that the benefits accrue in different spatial and temporal dimensions. Certain benefits accrue to local communities and populations, e.g. the subsistence use of timber and non-timber forest products. Other benefits, such as the erosion control and watershed protection functions, accrue regionally or nationally. Benefits from carbon sequestration and protection of biodiversity are essentially global in scale.

The carbon sequestration benefits of forests may be comparatively insignificant compared to the other functional and non-use values of tropical forests. As a first estimation of some of these benefits, for example, the option value of the potential globally commercial pharmaceutical products derived from forest plants per hectare may be in the region of very low to as much as US$20 per ha (Pearce and Puroshothaman, 1992). Although it is recognised that such an analysis can only demonstrate a small part of the value of biological resources such as forests (see Brown and Moran, 1994, for a discussion), it may provide a lower-bound estimate of a value hitherto unacknowledged as being of economic importance. Furthermore, a number of studies have indicated that the economic value of sustainably harvested timber and the non-timber forest products of tropical forests may be considerably greater than the gain from clear-cutting for timber production (Peters et al., 1989; Ruitenbeek, 1990). Although economic prescriptions for the conservation of rainforests may use their option value or carbon storage value as justification, successful prescriptions will be dependent on mechanisms for the increased local capture of the full range of benefits.

The benefits of *afforestation* schemes are of a different profile. The carbon sequestration profile of new afforestation is steeper in the first instance, depending on previous land use, climatic conditions and species planted. The local economic and environmental effects of afforestation depend on whether this is undertaken through plantation monoculture afforestation or through sustainable agroforestry. Related to the sovereignty issue is that of whether the forestry projects are in themselves desirable on social grounds, and whether the implicit shifting of property rights from local rights to global benefits leads to the undervaluing of these local benefits in the decision-making process.

8.6 DRAWBACKS OF INTERNATIONAL OFFSETS IN THE TROPICAL FOREST CONTEXT

The previous section has shown that there are environmental benefits to tropical forest conservation which are global and may be captured by the tropical countries through carbon offsets. However, several factors militate against these agreements. These include the additional transaction costs of potential forestry agreements compared to energy sector offsets; and the political debate as to the historic responsibility for greenhouse gas concentrations in the atmosphere from biological emissions.

Transaction costs

Transaction costs are the additional costs entailed in information gathering and drawing up of contracts in markets, and they fall on the party that does *not* retain the property rights. In the case of bilateral carbon offsets, the transaction costs would be incurred by the industrialised country attempting to offset its emissions in a tropical country. Every market transaction involves these costs in some form: from choosing food in a supermarket to purchasing commodities on international markets. Transaction costs fall on the party that does not have the property rights, i.e. the buyer. In the case of an international tradeable permit system for greenhouse gas emissions, an international agency would be required to facilitate the market transactions and also to monitor and regulate the market (Tietenberg, 1992).

Transaction costs are likely to be higher for bilateral offsets through forest conservation, than for the alternative types of offsets (in electricity generation technology, for example), because of the range of scientific uncertainty in carbon sequestration rates; the high unit costs of monitoring the proposed projects, even if these are large projects; and the potentially high cost of enforcement. The monitoring and enforcement of the agreement is essential, possibly by an international agency if international offsets become common.

Carbon sequestration rates from changing land use in tropical regions are highly uncertain. The range of carbon stored per hectare is dependent on different definitions of forest types; on assumptions of soil fluxes and characteristics; and on the previous and subsequent land use of the forested area (see Chapter 3 for a review). Large ranges exist in the estimates of carbon storage shown in Table 8.7 for different types of forest. Biomass and soil carbon levels after deforestation are also dependent on the management regime employed. The ranges are large for carbon storage in biomass and soils after change to agricultural regimes (e.g. in the lower part of Table 8.7), even with assumptions incorporated in the analysis of fluxes which simplify the true range (Brown, 1992).

Monitoring the costs of forestry conservation projects, where the aim is carbon sequestration and storage, would also be high. In the first place, there is considerable uncertainty over the extent and rate of current deforestation, and therefore monitoring conservation itself is likely to be extremely problematic. Grainger (1993) presents a comparison of various estimates made from 1970 to 1990 which highlights serious discrepancies. First, it is not clear whether estimates refer to the same definitions of deforestation—be it total clearance, selective logging, or degradation—or also different types and categories of forest. Second, methodologies and techniques of measurement differ, and remote sensing has done little to provide more accurate data. Remote-sensing techniques offer conflicting estimates, depending on resolution and sampling coverage. For example, two estimates of deforestation in Brazilian Amazonia using remote sensing provide very different figures. A group in the Brazilian National Space Agency (INPE) estimates a rate of deforestation at 1.7 million ha year^{-1} for 1978–1988, whereas another group at the same agency estimated 8.1 million ha year^{-1} for 1987. The different estimates were produced as a result of using different resolutions, and the second measured smoke from fires as an indicator of deforestation. Accurate estimates wholly based on remote-sensing measurement are not currently available, so on-the-ground monitoring of forest areas or even of stand growth would presently be the only option.

Table 8.7   Ranges of carbon storage potential in tropical land uses

	Total (t C ha^{-1})	Percentage range
Closed secondary forest	152–237	56
Forest fallow (closed)	121–136	12
Forest fallow (open)	50–56	12
Shifting cultivation (year 1)	41–92	125
Shifting cultivation (year 2)	47–111	136
Permanent cultivation	56–70	25
Pasture	46–80	74

Source: Based on Brown (1992), German Bundestag (1990), Houghton et al. (1987).
Notes: Assumes carbon will reach minimum after 5 years in cropland, after 2 years in pasture.

The largest problem in forest conservation for carbon offsets, however, is enforcement. Enforcement of forest conservation projects is notoriously difficult for a number of reasons. These include, in most countries, the lack of trained and motivated personnel and poor infrastructure (Grainger, 1993). But most significant is the failure to take account of local needs and local distribution of costs and benefits, unless projects can be set up where side payments are paid on a basis which ensures self-regulation by local individuals or groups of individuals.

Enhancing forest conservation also avoids some of the problems associated with large-scale afforestation in tropical countries (these include environmental problems such as decreased biodiversity, introduction of exotic species including pests, and also adverse social effects). Barnett (1992) asserts that such programmes pose threats to rural subsistence agriculturalists, both in terms of loss of traditional land tenure rights, and to the productivity of traditional agricultural systems. In addition, Barnett claims that given past performance of large-scale plantations schemes, it is questionable whether infrastructure, experience or political will exist to implement large-scale afforestation for carbon sequestration. The property rights regime associated with existing forests may be individual or communal in nature, depending on the cultural context and the objectives of the conservation. International financing for carbon offsets may require some change in the property rights for the management of the resource locally. However, the property rights to the carbon stored in the forest are also fundamentally changed and this constitutes a major objection of developing countries in terms of the allocation of responsibility for global warming, as discussed above.

Property rights and historic responsibility

International carbon offset deals have particular implications for the apportioning and attenuation of property rights concerning carbon sequestration. In the case of bilateral transfers, the country receiving the resources would have to agree to the emissions reduction being credited to the donor country for the system to be effective. This is implicitly giving up part of the property rights of the carbon sink—in effect, trading its carbon sequestration properties. In the negotiations leading to the drafting of the Climate Change Convention, developing countries took the stance that they had an unalienable right to exploit resources within their borders in a sustainable manner (however defined) (see Hyder, 1992, for a summary). Indeed this concept is enshrined in the Climate Change Convention: 'States have . . . the sovereign right to exploit their own resources pursuant to their own environmental and development policies' (preamble to Climate Change Convention).

It is clear that the 'selling' of carbon sequestration rights may not be acceptable to some tropical countries on the grounds that this represents a loss of sovereignty. In summary, the management, enforcement, and historic responsibility arguments make international carbon offsets in the forestry sector complex and politically sensitive undertakings which may involve high investment and implementation risks.

8.7 CONCLUSIONS

There are theoretical benefits to carbon offset agreements whether they are implemented as private agreements, as part of bilateral arrangements or multilateral funding initiatives for joint implementation of the Climate Change Convention. The financial costs of initiating forestry offsets in developing countries may be attractive and more cost-effective in terms of tonnes of CO_2 emissions offset than domestic abatement. However, a number of issues may make the implementation of such schemes problematic: transaction costs; enforcement and verification; clarification of property rights; and controversy over historic responsibility for global warming.

In addition, the scale of the emissions problem could not be solved by afforestation or stopping deforestation, even if the hurdles outlined here could be overcome. Bekkering's (1992) analysis of future global trends in greenhouse gas emissions has shown that reducing emissions from fossil fuels will have the greatest effect on atmospheric carbon concentrations between 1990 and 2100. Preventing deforestation could have the next biggest impact, Bekkering's calculations postulating that halting deforestation could reduce the expected carbon concentration in the atmosphere by 12 percent. Massive reforestation, of 500 million ha, plus the regeneration of 385 million ha of degraded forest, could reduce concentrations by a maximum of 11 percent (see also Chapter 7). The carbon sequestration estimates are based on reforestation taking place at a rate of 10 million ha a year, as estimated by Houghton (1990). However, the data for land available in tropical countries for reforestation are disputed by Barnett (1992) who claims that they fail to take account of ownership patterns, cost, and the ability of degraded land to grow trees. Doubts have been expressed from many quarters concerning the likelihood of implementing such plans, and clearly a more accurate estimate of the scale of feasible afforestation is necessary.

However, international financing for tropical forestry conservation is abundantly available from both national and international agencies, and from non-government organisations and even private companies. The objectives of this financing are diverse, so some carbon sequestration and storage may be unintentional secondary benefits. The conservation and sustainable use of *existing forests* bring about the greatest benefits (Adger et al., 1994). Such management ensures multiple benefits of tropical forests are maintained and enhanced. These include local use benefits accruing to local communities, soil erosion and watershed protection accruing locally, nationally and regionally, and the global benefits of both carbon storage (and carbon emissions avoided), and the protection of biodiversity. Research initiatives which focus on optimising these multiple benefits and operationalising such conservation policies are necessary. Clearly more effective management of the current forest estate should be given priority in terms of public policy and funding.

The other international policy instruments described in this chapter—international tradeable permits and carbon tax—are not presently on the agenda for implementation. It was an enormous feat to develop a text which everyone agreed upon and ensured that over 160 countries signed the Convention. Ambitious targets and commitments had to be sacrificed to ensure even the present agreement. Internationally co-ordinated action could promote cost-effective emission reductions, but would require that the system be seen to be fair, and that all major parties perceive additional advantages, rather than simply a reduced risk of climate change impacts, for co-operation to come about.

In the meantime, carbon offset deals are becoming increasingly common. The impetus of the Climate Change Convention may be halted if the Conference of the Parties is persuaded that joint implementation is unworkable. Indeed, joint implementation is a major stumbling block in the post-Convention negotiations of the international negotiating committee for the reasons outlined in this chapter. Several ways to make the concept acceptable, including limiting joint implementation to a maximum of 10 or 20 percent of a country's abatement effort, or giving credit for only a proportion of the emissions offset, have been suggested. The other radical proposal put forward by a Danish delegation to the International Negotiating Committee of the Convention is to exclude sink enhancement projects from the scheme altogether (Loske and Oberthür, 1994).

In the forest sector the problems identified in this chapter from the emitter's perspective of uncertainty and ultimately of a limited sink, and from the recipient's perspective of the ecological and social implications of forest management for maximum carbon sequestration, will limit forest offsets as a strategy. Ultimately joint implementation is not a panacea for least-cost climate stabilisation and will only form a small part of an overall strategy.

CHAPTER 9

Land Use Options for Greenhouse Gas Abatement: Prospects and Constraints

9.1 INTRODUCTION

This volume has set out methodologies for estimating the impact of land use and land use changes on the processes which enhance the greenhouse effect and which will result in global warming. We have presented examples of how changes of land use and its effects may be appraised, examined and measured in the light of scientific and economic uncertainty. In Part III of the book we have examined the driving forces of land use change processes contributing to the greenhouse effect and possible mitigation strategies from a social science perspective. In this concluding chapter we briefly review the contribution of land use and land use change to global warming, and the policy options for abatement strategies in the land use sector. The themes identified in the preceding chapters are discussed, and we summarise what we consider to be the issues most pertinent to future policy.

9.2 EMISSIONS OF GREENHOUSE GASES FROM LAND USE

In aggregate the emissions from land use sources are 20–30 percent of the annual emissions to the atmosphere across all gases, though this figure varies from country to country and depends on which natural sources and sinks are included. The world annual aggregate estimates of Edmonds et al. (1992) show that land use related activities contribute 31 percent to the global total. If a shorter time horizon is used to calculate the relative contribution of the gases this results in the contribution of land use activities being estimated as higher, due to the predominance of CH_4 and N_2O emissions in the total. This is a significant proportion of overall emissions, and policies to reduce these emissions should not be overlooked. Of these land use emissions, approximately half of CO_2 emissions and approximately 10 percent of CH_4 emissions are associated with forest loss (see Tables 1.1 and 3.15) in the most recent decade. Rice production, livestock and intensive agricultural practices as well as natural wetlands and other natural sources constitute the remaining land use related emissions.

The majority of the biological emissions are associated with human interference with the 'natural' system. Land use *change* is the principal human intervention. This has been illustrated for the UK, where the 66 percent of the land remaining in the same use over the last half century accumulated carbon (2.2 mt C year^{-1}), whereas the 34 percent of land which had changed its use caused a net emission of carbon (0.9 mt C year^{-1}). This net emission associated with land use change occurred despite net afforestation in the UK: coniferous afforestation actually added to net emissions through being carried out at the expense of previously mature broadleaved forest, upland heath, or on peatbogs such that carbon was instantaneously released.

Similarly on a global scale, deforestation, the changing of land out of forest, is the single most important land use related cause of the increase in atmospheric concentrations of CO_2. It is thought that the rate of deforestation has slowed since its peak in the 1980s, though the decadal global assessments carried out by the FAO do not allow a detailed picture of an issue complicated by seasonal weather variations, natural fires, and policy changes. The presumed slow-down has occurred in some countries (e.g. Brazil) as a result of changes of policies which previously encouraged deforestation, either inadvertently or as part of a development strategy, for political and economic reasons.

The rates of deforestation have multiple causes and explanations: the role of logging and concession policies, international indebtedness and shifting agricultural cultivation have variously been apportioned blame. Chapter 5 showed that some of these factors have indirect rather than direct relationships with the rate of deforestation in a particular country. The subsequent land use influences the impact of deforestation on the greenhouse effect. Burning of significant areas of forest brings about increased greenhouse impact through the emission of CH_4 and N_2O, impacts which do not occur if timber is harvested as its products subsequently store carbon for significant periods. However, all loss of forest area has numerous environmental and economic impacts: from displacement of indigenous people, to the loss of ecological functions of forests as protectors of soil and regulators of watersheds and as key habitats for biological diversity.

So the range of ecological and economic disturbance caused by much land use change is reinforced, as a rationale for conservation, by the impact of many land use changes on the net emission of greenhouse gases. This is not always the case, however, as the management of natural wetlands is an example of how human intervention actually reduces instantaneous emissions of CH_4. The case has even been made for natural wetland drainage (Bouwman, 1990c, p. xvii) for this reason. The net effect of the historical reduction of the area of natural wetlands through human intervention does not support this viewpoint, as in the long term, the carbon accumulation in the anaerobic conditions of wetlands is the overiding factor, the net impact of increased wetland drainage is increased atmospheric concentrations of carbon. But this example illustrates the trade-offs implicit in these land use management decisions.

Overall, the assessment of the sources of land use related greenhouse gases shows that in general a precautionary approach to land use change postpones the onset and reduces the potential magnitude of climate changes associated with the enhanced greenhouse effect. The precautionary principle asserts that in the face of uncertainty action should be taken which gives 'breathing space' before more information or better technology becomes available. To state the principle in economic terms, decision-makers are recommended to take precautionary action where the expected value of a cost or benefit cannot be calculated without fuller knowledge of the probability distribution and a cost benefit calculation is inappropriate (Perrings, 1991).

In the case of land use related greenhouse gas emissions the issue of whether to take precautionary action is not easily reconciled, as those decision-makers who are concerned with future climate change are often not those who take direct land use decisions. The issue of whether action on land use related impacts is desirable for governments, land users and decision-makers at each level is the main thrust of Part III of this book. The greenhouse effect cannot be isolated from other economic and ecological factors.

9.3 ABATEMENT OPTIONS

Part III has analysed a number of issues surrounding the causes of, and options for reducing, greenhouse gas emissions from land use. The causes of deforestation illustrate how social and economic factors influence land use: deforestation is caused by land users responding to economic and social signals which have deleterious consequences. Options for changing present greenhouse gas emissions lie either in influencing land use change (halting deforestation, preserving existing carbon sinks), or in changing management of land (reducing emissions from livestock, or from rice growing). Various themes have resulted from particular case studies examined. These are summarised below.

The nature of the global environmental problem

This book has approached the issue of land use related greenhouse gas emissions as a public goods problem associated with the nature of the atmosphere as a global open access resource. Policies to reduce emissions of the gases in the land use sector are hampered by distanciation from the impacts of global warming. People who make land use decisions are often separated in both space and time from the consequences of their actions. The most prodigious emitters of greenhouse gases are not the people who will suffer most severely the effects of climate change. In the first instance, the rich industrialised countries, the largest emitters both historically and currently, will be more able to adapt to the impacts of climate change such as rising sea level and other environmental hazards. Secondly, the impacts of global warming will not be felt to full effect for some decades. This means that even if the most accurate and up-to-date information on climate change is brought to present emitters, they may not have sufficient incentives to undertake precautionary action, because the seemingly 'rational' course of action may be to ignore this future problem.

One mechanism by which this global commons problem can be overcome is through co-ordinated international action and agreement. The Climate Change Convention, signed in 1992 by over 160 countries entered into force in March 1994 (through being ratified by more than 50 signatories). It sought as its guiding principle to take the precautionary action required to avoid dangerous anthropogenic interference with the climate system. The target set, of stabilising emissions in all countries at 1990 levels by the year 2000, will not achieve this stated objective. Reaching this target will not avoid climate change altogether; some impacts are already committed to. Whether the mechanism of the Convention will set new, more stringent targets beyond the year 2000, remains to be seen. Such measures will be necessary if the world is to avoid severe climatic disruption, and the accompanying physical, social and economic impacts.

Opportunities for cost-effective abatement do exist

Some technical options for abatement of greenhouse gases from land use sources have been identified, particularly in Part III of the book. Some of these then are win–win options, in that the rationale exists for the actions to be undertaken *even without consideration of the greenhouse gas implications*. Options explored in Chapters 5–8 include:

— The reduction of deforestation: an option with multiple environmental and economic benefits.
— Management options in rice production: research is required which increases the ratio of productivity of rice production to emissions of CH_4. There is little scope however for reducing growing aggregate demand for rice over time, given that it is the staple food of half the world's population.
— Livestock numbers reduction and management changes: Chapter 6 showed that the monetary value of the methane emissions avoided (as estimated by the global warming damage avoided) is a significant factor (up to 40 percent) in offsetting the cost of reducing support to the livestock sector in the UK. This is without consideration of the other environmental and structural implications of changing support on the agricultural sector.
— Afforestation to sequester and store carbon: in the UK example the costs may be £10 per t C sequestered. However, afforestation as an option requires careful consideration. First, because monoculture plantations may have detrimental ecological and social impacts in both temperate and tropical

situations. Secondly, the assumption of 'surplus' land with zero, or very low, opportunity cost, may be erroneous.

Policy and market failures and perverse regulations create barriers to the implementation of emission reduction policies. But there is also a profound lack of information about these cost-effective abatement strategies. Economists may assume that these options are non-feasible, without ever giving them full consideration, just as:

> the econometrician who, asked by his mannerly granddaughter whether she could pick up a $20 bill she'd just noticed lying on the sidewalk, replied 'No dear, don't bother: if it were real, someone would have picked it up already'.
>
> (Lovins and Lovins, 1992, pp. 433–434)

If some or all of the options for abatement were taken up this would produce a global benefit in terms of climate change avoided. Some of the options identified would have resource costs, and these costs should be fairly distributed. Neither the 'polluter pays principle', nor the 'victim pays principle' is acceptable in the case of land use related emissions of greenhouse gases, however. Blame cannot be apportioned along such simplistic lines: the responsibility for historic emissions remains a moot point, and because of the global commons nature of the atmosphere and climate system, everyone and each nation is a polluter and a victim, to varying degrees.

International transfers of resources

The Convention on Climate Change explicitly stated that the signatory parties had common responsibility for the global warming problem, but also that the extent of the responsibility is 'differentiated' between countries. The countries which are presently industrialised, who have historically been responsible for greenhouse gas emissions through the burning of fossil fuel and through deforestation and land use change, should now take on the responsibility for funding both emissions reduction and the costs of adaptation to climate change in the poorer countries. Although the signatory parties may have been well intentioned in signing the Convention, no significant transfers of resources have been pledged in the following two years. This is partly as a result of political difficulties concerning the operation of the GEF, designated as the manager of the financial mechanism under the Convention (Jordan, 1994); and also because of a lack of political will to deal in a precautionary manner with a global environmental problem of uncertain magnitude.

Various mechanisms were suggested in the lead up to the signing of the Convention that would help countries meet the targets of the Convention with least resource cost. These include internationally administered carbon taxes or tradeable permits. Although theoretically elegant, such proposals are unlikely to become reality due to their inevitable conflict with national interests in some countries, and due to the institutional hurdles of implementation. The difficulties experienced in dispensing the US$1.1 billion under the pilot phase of the GEF could be multiplied considerably in a system with more money and more numerous countries. Joint implementation of the Convention, where one country offsets its domestic emissions through financing emission reduction or sink enhancement in another country, is also theoretically a cost-saving mechanism. Unlike the proposals for tradeable permits and taxes, offset deals are presently being funded in Central America, Russia and elsewhere, though their validity under the Climate Change Convention itself remains open to question. Afforestation projects are widely discussed in this context: the advantages are that funding allows the financing of forest conservation while the funder claims credit for the carbon locked up in the biomass. However, Chapter 8 has shown that offset projects associated with biological emissions are unlikely to be undertaken because of the high levels of uncertainty in the carbon sequestration potential. Even if the potential carbon sequestration associated with offset deals could be accurately identified, the implementation of forest conservation is a fundamental problem. It is difficult to design institutions which provide forest conservation if the existing economic and policy environment does not promote sustainable forest use, so any projects would need to be supported by a policy environment which reverses incentives to degrade and destroy forests.

9.4 CONCLUSION

It is possible to design land use strategies that will reduce emissions of greenhouse gases. These appear to be technically and economically feasible, although current policy failures may militate against their implementation. These distortions must be corrected in order to ensure that more sustainable forms of land use are undertaken, which minimise emissions of greenhouse gases and other kinds of environmental degradation. Decisions about different land uses inevitably involve trade-offs, and economic analysis can assist in assessing the effects of different policies. In the case of global warming, trade-offs between present and future welfare may be necessary and such decisions may be particularly hard for politicians to make. Many of the options for reduction of greenhouse gas emissions in the land use sector which we have suggested in this volume are win–win policies, which may bring additional benefits. One example is the conservation of existing forests, as compared with new afforestation schemes. Such policies will be more attractive, particularly for developing countries, where trade-offs between environment and development may be more acute.

The Climate Change Convention provides a framework for international co-operation, under which strategies for sustainable land use can be developed and implemented, perhaps involving international transfers of finance and technology. Current targets will not avoid impacts of climate change, but they provide a focal point for international action. In compiling inventories of greenhouse gas sinks and sources, parties to the Convention will have to explicitly consider the land use sector as part of their response to the global threat of climate change. This is a necessary, though not sufficient, step towards sustainable development in the land use sector.

References

Adams, R.M., Chang, C.-C., McCarl, B.A. and Callaway, J.M. (1992), 'The role of agriculture in climate change: a preliminary evaluation of emission-control strategies', in J.M. Reilly and M. Anderson (eds), *Economic Issues in Global Climate Change: Agriculture, Forestry and Natural Resources*, Boulder, Co, Westview Press, pp. 273–287.

Adams, W.M. (1992), *Green Development: Environment and Sustainability in the Third World*, London, Routledge.

Adger, W.N. (1993), 'Agriculture and the environment', in D.W. Pearce (ed.), *Blueprint 3: Measuring Sustainable Development*, London, Earthscan, pp. 115–130.

Adger, W.N. and Brown, K. (1993), 'A UK greenhouse gas inventory: on estimating anthropogenic and natural sources and sinks', *Ambio*, **22**: 509–517.

Adger, W.N. and Fankhauser, S. (1993), 'Economic analysis of the greenhouse effect: optimal abatement level and strategies for mitigation', *International Journal of Environment and Pollution*, **3**: 104–119.

Adger, W.N. and Moran, D.C. (1993), 'Estimating the benefits of greenhouse gas emission reduction from agricultural policy reform', *World Resource Review*, **5**: 303–323.

Adger, W.N., Brown, K., Cervigni, R. and Moran, D. (1994), *Towards Estimating Total Economic Value of Forests in Mexico*, Global Environmental Change Working Paper 94–21, Centre for Social and Economic Research on the Global Environment, University of East Anglia, University College London.

Adger, W.N., Brown, K., Shiel, R.S. and Whitby, M.C. (1992a), 'Carbon dynamics of land use in Great Britain', *Journal of Environmental Management*, **36**: 117–133.

Adger, W.N., Brown, K., Shiel, R.S. and Whitby, M.C. (1992b), 'Sequestrations and emissions from agriculture and forestry: carbon in the dock', *Land Use Policy*, **9**: 122–130.

Agarwal, A. and Narain, S. (1991), *Global Warming in an Unequal World: A Case of Environmental Colonialism*, New Delhi, Centre for Science and Environment.

Ahmed, S. (1993), 'The role of agriculture in the economies of some Asian-Pacific countries: an overview', in S. Ahmed and A.L. Clark (eds), *Fertiliser Policy in Asia and the Pacific*, Honolulu, Asian Productivity Organisation and East–West Centre, pp. 11–19.

Ahuja, D.R. (1992), 'Estimating national contributions of greenhouse gas emissions: the CSE-WRI controversy', *Global Environmental Change*, **2**: 83–87.

Allen, J. and Barnes, D. (1985), 'The causes of deforestation in developing countries', *Annals of the Association of American Geographers*, **75**: 163–184.

Anderson, A.B. (ed.) (1990), *Alternatives to Deforestation*, New York, Columbia University Press.

Anderson, D. (1987), *The Economics of Afforestation: A Case Study in Africa*, Baltimore,

Johns Hopkins University Press for the World Bank.

Anderson, D. (1991), *The Forestry Industry and the Greenhouse Effect*, Edinburgh, Forestry Commission and the Scottish Forestry Trust.

Andrasko, K. (1990), 'Global warming and forests: an overview of current knowledge', *Unasylva*, **40**: 3–11.

Armentano, T.V. and Menges, E.S. (1986), 'Patterns of change in the carbon balance of organic soil wetlands of the temperate zone', *Journal of Ecology*, **74**: 755–774.

Aselmann, I. and Crutzen, P.J. (1989), 'Global distribution of natural freshwater wetlands and rice paddies, their net primary productivity, seasonality and possible methane emissions', *Journal of Atmospheric Chemistry*, **8**: 307–358.

Askew, G.P., Payton, R.W. and Shiel, R.S. (1985), 'Upland soils and land clearance in Britain during the second millennium BC', in D. Spratt and C. Burgess (eds), *Upland Settlement in Britain: The Second Millennium BC and After*, London, British Archaeological Report British Series 143, pp. 5–33.

Avery, B.W. (1980), *Soil Classification for England and Wales*, Soil Survey Technical Monograph 14, Soil Survey of England and Wales, Harpenden.

Ayres, R.U. and Walter, J. (1991), 'The greenhouse effect: damages, costs and abatement', *Environmental and Resource Economics*, **1**: 237–270.

Bachelet, D. and Neue, H.U. (1993), 'Methane emissions from wetland rice areas of Asia', *Chemosphere*, **26**: 219–237.

Barbier, E.B. and Pearce, D.W. (1990), 'Thinking economically about climate change', *Energy Policy*, **18**: 11–18.

Barker, T. (1993), 'The carbon tax: economic and policy issues', in C. Carraro and D. Siniscalco (eds), *The European Carbon Tax: An Economic Assessment*, Dordrecht, Kluwer, pp. 239–254.

Barnett, A. (1992), *Deserts of Trees: the Environmental and Social Impacts of Large-scale Tropical Reforestation in Response to Global Climate Change*, London, Friends of the Earth.

Barrett, S. (1991), 'Global warming: economics of a carbon tax', in D.W. Pearce (ed.), *Blueprint 2: Greening the World Economy*, London, Earthscan, pp. 31–52.

Barrett, S. (1992), 'Acceptable allocations of tradeable carbon emission entitlements in a global warming treaty', in UNCTAD (ed.), *Combating Global Warming: Study on a Global System of Tradeable Carbon Emission Entitlements*, New York, United Nations, pp. 85–113.

Bartlett, K.B. and Harriss, R.C. (1993), 'Review and assessment of methane emissions from wetlands', *Chemosphere*, **26**: 261–320.

Bartlett, K.B., Crill, P.M., Bonassi, J.A., Richey, J.E. and Harriss, R.C. (1990), 'Methane flux from the Amazon River floodplain: emissions during rising water', *Journal of Geophysical Research*, **95**: 16 773–16 788.

Baumol, W.J. and Oates, W.E. (1988), *The Theory of Environmental Policy*, 2nd edn, Cambridge, Cambridge University Press.

Bazzaz, F.A. and Fajer, E.D. (1992), 'Plant life in a CO_2 rich world', *Scientific American*, **266** (January): 18–24.

Bekkering, T.D. (1992), 'Using tropical forests to fix atmospheric carbon: the potential in theory and practice', *Ambio*, **21**: 414–419.

Bennett, A.J. (1992), 'The dynamics of agriculture and forestry at the forest edge', in F.R. Miller and K.L. Adam (eds), *Oxford Conference on Tropical Forests*, Oxford, Oxford Forestry Institute, pp. 139–148.

Bertram, G. (1992), 'Tradeable emission permits and the control of greenhouse gases', *Journal of Development Studies*, **28**: 423–446.

Bilsborrow, R.E. and Goeres, M. (1994), 'Population, land use and the environment in developing countries: what can we learn from cross-national data?', in K. Brown and D.W. Pearce (eds), *The Causes of Tropical Deforestation*, London, UCL Press, pp. 106–133.

Bilsborrow, R.E. and Okoth Ogendo, H.W.O. (1992), 'Population driven changes in land use in developing countries', *Ambio*, **21**: 37–45.

Birdsall, N. and Steer, A. (1993), 'Act now on global warming—but don't cook the books', *Finance and Development*, **30** (1): 6–8.

Bishop, K. (1992), 'Assessing the benefits of community forests: an evaluation of the recreational use benefits of two urban fringe woodlands', *Journal of Environmental Planning and Management*, **35**: 63–76.

Blum, E. (1993), 'Making biodiversity conservation profitable: a case study of the Merck/INBio agreement', *Environment*, **35** (April): 16–20, 38–45.

Boden, T.A., Sepanski, R.J. and Stoss, F.W. (1991), *Trends 91: A Compendium of Data on Global Change*, ORNL/CDIAC-46, Oak Ridge, TN, Oak Ridge National Laboratory.

Bolin, B., Döös, B.R., Jäger, J. and Warrick, R. (eds) (1986), *The Greenhouse Effect, Climate Change and Ecosystems*, Scope 29, Chichester, John Wiley, pp. 157–203.

Bolle, H.J., Seiler, W. and Bolin, B. (1986), 'Other greenhouse gases and aerosols', in B. Bolin, B.R. Döös, J. Jäger and R.A. Warrick (eds), *The Greenhouse Effect, Climate Change and Ecosystems*, Scope 29, Chichester, John Wiley, pp. 157–203.

Boserup, E. (1965), *The Conditions of Agricultural Growth: The Economics of Agrarian Change under Population Pressure*, London, Allen and Unwin.

Bouwman, A.F. (1990a), 'Exchange of greenhouse gases between terrestrial ecosystems and the atmosphere', in A.F. Bouwman (ed.), *Soils and the Greenhouse Effect*, Chichester, John Wiley, pp. 61–127.

Bouwman, A.F. (1990b), 'Land use related sources of greenhouse gases: present emissions and possible future trends', *Land Use Policy*, **7**: 154–164.

Bouwman, A.F. (ed.) (1990c), *Soils and the Greenhouse Effect*, Chichester, John Wiley.

Bouwman, A.F., Fung, I., Matthews, E. and John, J. (1993), 'Global analysis of the potential for N_2O production in natural soils', *Global Biogeochemical Cycles*, **7**: 557–597.

Brown, K. (1985), *Labour Saving Technologies for Rural Women in LDCs: A Study of Woodstove Innovations*, MSc Thesis, Tropical Agricultural Development, University of Reading.

Brown, K. (1992), *Carbon Sequestration and Storage in Tropical Forests*, Global Environmental Change Working Paper 92-24, Centre for Social and Economic Research on the Global Environment, University of East Anglia, University College London.

Brown, K. and Adger, W.N. (1993), 'Estimating national greenhouse gas emissions under the Climate Change Convention', *Global Environmental Change*, **3**: 149–158.

Brown, K. and Adger, W.N. (1994), 'Economic and political feasibility of international carbon offsets', *Forest Ecology and Management*, **68**.

Brown, K. and Maddison, D. (1993), 'The UK and the global environment', in D.W. Pearce (ed.), *Blueprint 3: Measuring Sustainable Development*, London, Earthscan, pp. 167–182.

Brown, K. and Moran, D. (1994), 'Valuing biodiversity: the scope and limitations of economic analysis', in V. Sánchez and C. Juma (eds), *Biodiplomacy: Genetic Resources and International Relations*, Nairobi, ACTS Press, pp. 213–232.

Brown, K. and Pearce, D.W. (eds) (1994a), *The Causes of Tropical Deforestation*, London, UCL Press.

Brown, K. and Pearce, D.W. (1994b), 'The economic value of non-marketed benefits of tropical forests: carbon storage', in J. Weiss (ed.), *The Economics of Project Appraisal and the Environment*, London, Edward Elgar, pp. 102–123.

Brown, K., Adger, W.N. and Turner, R.K. (1993), 'Global environmental change and mechanisms for North–South resource transfers', *Journal of International Development*, 5: 571–589.

Brown, S. and Lugo, A.E. (1984), 'Biomass of tropical forests: a new estimate based on forest volumes', *Science*, 223: 1290–1293.

Brown, S. and Lugo, A.E. (1992), 'Above ground biomass estimates for tropical moist forests of the Brazilian Amazon', *Interciencia*, 17: 8–18.

Brown, S., Lugo, A.E. and Wisniewski, J. (1992), 'Missing carbon dioxide', *Science*, 257: 11.

Bruijneel, L.A. (1992), 'Managing tropical forest watersheds for production: where contradictory theory and practice co-exist', in F.R. Miller and K.L. Adam (eds), *Oxford Conference on Tropical Forests*, Oxford, Oxford Forestry Institute, pp. 37–76.

Buckwell, A.E. (1991), 'The CAP and world trade', in C. Ritson and D. Harvey (eds), *The Common Agricultural Policy and the World Economy: Essays in Honour of John Ashton*, Wallingford, CAB International, pp. 223–240.

Buckwell, A.E. (1992), 'Should we set aside set-aside?', in J. Clarke (ed.), *Set-Aside*, Farnham, Surrey, British Crop Protection Council, Monograph 50, pp. 275–283.

Bunce, R.G.H. (1993), *Terrestrial Vegetation: Composition, Distribution and Change*, Countryside Survey 1990: A Preview, Merlewood, Cumbria, Institute for Terrestrial Ecology.

Bunce, R.G.H., Barr, C.J. and Fuller, R.M. (1992), 'Integration of methods for detecting land use change with special reference to Countryside Survey 1990', in M.C. Whitby (ed.), *Land Use Change: The Causes and Consequences*, ITE Symposium No. 27, London, HMSO, pp. 69–78.

Burgess, J. (1991), *Economic Analysis of Frontier Agricultural Expansion and Tropical Deforestation*, MSc Thesis, Environmental and Resource Economics, University College London.

Burgess, J. (1992), *Economic Analysis of the Causes of Tropical Deforestation*, Discussion Paper 92-03, London Environmental Economics Centre.

Burke, M.K., Houghton, R.A. and Woodwell, G.M. (1990), 'Progress towards predicting the potential for increased emissions of CH_4 from wetlands as a consequence of global warming', in A.F. Bouwman (ed.), *Soils and the Greenhouse Effect*, Chichester, John Wiley, pp. 451–455.

Buttel, F. (1993), 'Environmentalisation and greening: origins, processes and implications', in S. Harper (ed.), *The Greening of Rural Policy: International Perspectives*, London, Belhaven, pp. 12–26.

Cannell, M.G.R. (1982), *World Forest Biomass and Primary Production Data*, London, Academic Press.

Cannell, M.G.R. and Cape, J.N. (1991), 'International environment considerations: acid rain and the greenhouse effect', in Forestry Commission (ed.), *Forestry Expansion: a Case Study of Technical, Economic and Ecological Factors*, Edinburgh, Forestry Commission.

Capistrano, D. and Kiker, C. (1990), *Global Economic Influences on Tropical Broadleaved Forest Depletion*, Washington, DC, World Bank.

Carraro, C. and Siniscalco, D. (eds) (1993), *The European Carbon Tax: An Economic Assessment*, Dordrecht, Kluwer.

Cheng, Y.S. (1984), 'Effects of drainage on the characteristics of paddy soils in China', in IRRI (eds), *Organic Matter and Rice*, Los Baños, Philippines, International Rice Research Institute, pp. 417–426.

Cicerone, R.J. and Oremland, R.S. (1988), 'Biogeochemical aspects of atmospheric methane', *Global Biogeochemical Cycles*, **2**: 299–327.

Cline, W.R. (1992), *The Economics of Global Warming*, Washington, DC, Institute for International Economics.

Cline, W.R. (1993), 'Give greenhouse abatement a fair chance', *Finance and Development*, **30(1)**: 3–5.

Clymo, R.S. (1983), 'Peat', in A.J.P. Gore (ed.), *Ecosystems of the World Vol. 4A: Mires; Swamp, Bog, Fen and Moor: General Studies*, Amsterdam, Elsevier, pp. 159–224.

Clymo, R.S. (1984), 'The limits to peat bog growth', *Philosophical Transactions of the Royal Society*, **303B**: 605–654.

Clymo, R.S. and Reddaway, E.J.F. (1971), 'Productivity of sphagnum (bog-moss) and peat accumulation', *Hydrobiologia*, **12**: 181–192.

Cofer, W.R., Levine, J.S., Winstead, E.L. and Stocks, B.J. (1991), 'Nitrous oxide emissions from biomass burning', *Nature*, **349**: 689–691.

Cofer, W.R., Levine, J.S., Winstead, E.L., Stocks, B.J., Cahoon, D.R. and Pinto, J.P. (1993), 'Trace gas emissions from tropical biomass fires: Yucatan Peninsula, Mexico', *Atmospheric Environment*, **27A**: 1903–1907.

Colchester, M. and Lohmann, L. (eds) (1993), *The Struggle for the Land and the Fate of the Forests*, London, Zed Books.

Cole, C.V., Flach, K., Lee, J., Sauerbeck, D. and Stewart, B. (1993), 'Agricultural sources and sinks of carbon', *Water, Air and Soil Pollution*, **70**: 111–122.

Constantino, L. and Ingram, D. (1990), *Supply–Demand Projections for the Indonesian Forestry Sector*, Jakarta, FAO.

Conway, G.R. and Pretty, J.N. (1991), *Unwelcome Harvest: Agriculture and Pollution*, London, Earthscan.

Cooke, G.W. (1967), *The Control of Soil Fertility*, London, Crosby Lockwood.

Cornia, G.A., van der Hoeven, R. and Mkandawire, T. (eds) (1992), *Africa's Recovery in the 1990s: from Stagnation and Adjustment to Human Development*, New York, St. Martin's Press.

Costigan, P. (1993), 'Methane emissions from UK agriculture', in A. Williams (ed.), *Methane Emissions*, London, Watt Committee on Energy, pp. 105–112.

Cox, P.A. and Elmqvist, T. (1991), 'Indigenous control of tropical rainforest reserves: an alternative strategy for conservation', *Ambio*, **20**: 317–321.

Crabtree, J.R. and Macmillan, D.C. (1989), 'UK fiscal changes and new forestry planting', *Journal of Agricultural Economics*, **40**: 314–322.

Crompton, E. (1953), 'Grow the soil to grow the grass: some pedological aspects of marginal land improvement', *Agriculture: Journal of the Ministry of Agriculture*, **60**: 301–308.

Cromwell, E. and Winpenny, J. (1993), 'Does economic reform harm the environment? a review of structural adjustment in Malawi', *Journal of International Development*, **5**: 623–650.

Crutzen, P.J. (1991), 'Methane's sinks and sources', *Nature*, **350**: 380–381.

Crutzen, P.J. and Andreae, M.O. (1990), 'Biomass burning in the tropics: impact on atmospheric chemistry and biogeochemical cycles', *Science*, **250**: 1669–1677.

Crutzen, P.J., Aselmann, I. and Seiler, W. (1986), 'Methane production by domestic animals, wild ruminants, other herbivorous fauna and humans', *Tellus*, **38B**: 271–284.

Crutzen, P.J., Heidt, L.E., Krasnec, J.P., Pollock, W.H. and Seiler, W. (1979), 'Biomass burning as a source of atmospheric gases CO, H_2, N_2O, NO, CH_3Cl and COS', *Nature*, **282**: 253–256.

Dai, A. and Fung, I.Y. (1993), 'Can climate variability contribute to the missing CO_2 sink', *Global Biogeochemical Cycles*, **7**: 599–609.

Daniels, R.C., White, T.W. and Chapman, K.K. (1993), 'Sea-level rise: destruction of threatened and endangered species habitat in South Carolina', *Environmental Management*, **17**: 373–385.

Davidson, J. (1985), 'Economic use of moist tropical forests', *The Environmentalist*, **9**: Supplement.

Deacon, R.T. and Murphy, P. (1992), *The Structure of an Environmental Transaction: The Debt-for-Nature Swap*, Annaheim, California, Allied Social Science Associations Meetings.

Denier van der Gon, H.A.C., Neue, H.U., Lantin, R.S., Wassmann, R., Alberto, M.C.R., Aduna, J.B. and Tan, M.J.P. (1992), 'Controlling factors of methane emission from rice fields', in N.H. Batjes and E.M. Bridges (eds), *World Inventory of Soil Emission Potentials*, Wageningen, Netherlands, Wageningen Agricultural University, pp. 81–92.

Department of Energy (1991), *Energy Paper 59: Energy Related Carbon Emissions in Possible Future Scenarios in the United Kingdom*, London, HMSO.

Department of Environment (1993), *Climate Change: Our National Programme for CO_2 Emissions*, London, HMSO.

Detwiler, R.P. and Hall, C.A.S. (1988), 'Tropical forests and the global carbon cycle', *Science*, **239**: 42–47.

Devol, A.H., Richey, J.E., Clark, W.A., King, S.L. and Martinelli, L.A. (1988), 'Methane emissions to the troposphere from the Amazon floodplain', *Journal of Geophysical Research*, **93**: 1583–1592.

Devol, A.H., Richey, J.E., Forsberg, B.R. and Martinelli, L.A. (1990), 'Seasonal dynamics in methane emissions from the Amazon River floodplain to the troposphere', *Journal of Geophysical Research*, **95**: 16 417–16 426.

Dewar, R.C. (1990), 'A model of carbon storage in trees and timber', *Tree Physiology*, **6**: 417–428.

Dewar, R.C. and Cannell, M.G.R. (1992), 'Carbon sequestration in the trees, products and soils of forest plantations: an analysis using UK examples', *Tree Physiology*, **11**: 49–71.

Dixon, R.K., Winjum, J.K. and Schroeder, P.E. (1993a), 'Conservation and sequestration of carbon: the potential of forest and agroforest management practices', *Global Environmental Change*, **3**: 159–173.

Dixon, R.K., Andrasko, K.J., Sussman, F.A., Lavinson, M.A., Trexler, M.C. and Vinson, T.S. (1993b), 'Forest sector carbon offset projects: near-term opportunities to mitigate greenhouse gas emissions', *Water, Air and Soil Pollution*, **70**: 561–577.

Drennen, T.E. (1993), 'After Rio: the status of climate change negotiations', in H.M. Kaiser and T.E. Drennen (eds), *Agricultural Dimensions of Global Climate Change*,

Delray Beach, FL, St. Lucie Press, pp. 198–213.

Drennen, T.E. and Chapman, D. (1992), 'Negotiating a response to climate change: role of biological emissions', *Contemporary Policy Issues*, **10**: 49–58.

Duxbury, J.M. and Mosier, A.R. (1993), 'Status and issues concerning agricultural emissions of greenhouse gases', in H.M. Kaiser and T.E. Drennen (eds), *Agricultural Dimensions of Global Climate Change*, Delray Beach, FL, St. Lucie Press, pp. 229–258.

Eckholm, E. (1976), *Losing Ground: Environmental Stress in Developing Countries*, Washington DC, Worldwatch Institute.

Eckholm, E., Foley, G., Barnard, G. and Timberlake, L. (1984), *Fuelwood: The Energy Crisis That Won't Go Away*, London, International Institute for Environment and Development.

Edmonds, J. (1992), 'Why understanding the natural sinks and sources of CO_2 is important: a policy analysis perspective', *Water, Air and Soil Pollution*, **64**: 11–21.

Edmonds, J., Callaway, J.M. and Barns, D. (1992), 'Agriculture in a comprehensive trace-gas strategy', in J.M. Reilly and M. Anderson (eds), *Economic Issues in Global Climate Change: Agriculture, Forestry and Natural Resources*, Boulder, Co., Westview Press, pp. 56–71.

Edwards, N.T. and Ross-Todd, B.M. (1983), 'Soil carbon dynamics in a mixed deciduous forest following clear cutting with and without residue removal', *Soil Science Society of America Journal*, **47**: 1014–1021.

Edwards, N.T., Shugart, H.H., McLaughlin, S.B., Harris, W.F. and Reichle, D.E. (1981), 'Carbon metabolism in terrestrial ecosystems', in D.E. Reichle (ed.), *Analysis of Temperate Forest Ecosystems*, London, Chapman and Hall.

Ehrlich, A. (1990), 'Agricultural contributions to global warming', in J. Leggett (ed.), *Global Warming: The Greenpeace Report*, Oxford, Oxford University Press, pp. 400–420.

Epstein, P.R., Ford, T.E. and Colwell, R.R. (1993), 'Marine ecosystems: health and climate change', *The Lancet*, **342**: 1216–1219.

Euroconsult (1989), *Agricultural Compendium for Rural Development in the Tropics and Subtropics*, 3rd Edn, Amsterdam, Elsevier.

Faeth, P., Cort, C. and Livernash, R. (1994), *Evaluating the Carbon Sequestration Benefits of Forestry Projects in Developing Countries*, Washington DC, World Resources Institute.

Fankhauser, S. (1992), *Global Warming Damage Costs: Some Monetary Estimates*, Global Environmental Change Working Paper 92-29, Centre for Social and Economic Research on the Global Environment, University of East Anglia and University College London.

Fankhauser, S. (1994), *Valuing Climate Change: the Economics of the Greenhouse Effect*, London, Earthscan..

FAO (1991), *World Soil Resources: an Explanatory Note on the FAO World Soil Resources Map*, Report No.66, Rome, Food and Agriculture Organisation.

FAO (1992), *Production Yearbook, 1991*, Rome, Food and Agriculture Organisation.

Fearnside, P.M. (1985), 'Brazil's Amazon forest and the global carbon problem', *Interciencia*, **10**: 179–186.

Fearnside, P.M. (1992a), 'Forest biomass in Brazilian Amazonia: comments on the estimate by Brown and Lugo', *Interciencia*, **17**: 19–27.

Fearnside, P.M. (1992b), *Greenhouse Gas Emissions from Deforestation in the Brazilian Amazon, Carbon Emissions and Sequestration in Forests: Case Studies from Developing Countries*, Report LBL 32758, Berkeley, CA, Lawrence Berkeley Laboratory.

Fearnside, P.M., Leal, N. and Moreira Fernandes, F. (1993), 'Rainforest burning and the global carbon budget: biomass, combustion efficiency and charcoal formation in the Brazilian Amazon', *Journal of Geophysical Research*, **98**: 16 733–16 743.

Flint, M. (1992), 'Biological diversity and developing countries', in A. Markandya and J. Richardson (eds), *The Earthscan Reader in Environmental Economics*, London, Earthscan, pp. 437–469.

Fortmann, L. and Bruce, J. (eds) (1988), *Whose Tree? Proprietary Dimensions of Forestry*, Boulder, CO, Westview Press.

Franken, R.O.G., van Vierssen, W. and Lubberding, H.J. (1992), 'Emissions of some greenhouse gases from aquatic and semi-aquatic ecosystems in the Netherlands and options to control them', *Science of the Total Environment*, **126**: 277–293.

Freedman, B., Meth, F. and Hickman, C. (1992), 'Temperate forests as a carbon storage reservoir for carbon dioxide emitted by coal-fired generating stations: a case study for New Brunswick, Canada', *Forest Ecology and Management*, **55**: 15–29.

Freeman, C., Lock, M.A. and Reynolds, B. (1993), 'Fluxes of CO_2, CH_4 and N_2O from a Welsh peatland following simulation of water-table draw-down: potential feed-back to climate change', *Biogeochemistry*, **19**: 51–60.

Fujii, Y. (1990), *An Assessment of the Responsibility for the Increase in Carbon Dioxide Concentration and Inter-generational Carbon Accounts*, Luxemburg, WP 90-55, International Institute for Applied Systems Analysis.

German Bundestag (1990), *Protecting the Tropical Forests: A High Priority International Task*, Bonn, Bonner Universitäts-Buchdruckerei.

Gibbs, M.J. and Woodbury, J.W. (1993), 'Methane emissions from livestock manure', in A.R. Van Amstel (ed.), *Methane and Nitrous Oxide: Methods in National Emissions Inventories and Options for Control: IPCC Workshop*, Bilthoven, RIVM, pp. 81–91.

Giles, C. and Ridge, M. (1993), 'The impact on households of the 1993 Budget and the Council Tax', *Fiscal Studies*, **14** (August): 1–20.

Global Environment Facility (1992a), *Norwegian Funding of Pilot Demonstration Projects for Joint Implementation Arrangements under the Climate Convention: Memorandum of Understanding*, GEF Memorandum, Washington DC, GEF.

Global Environment Facility (1992b), *Report by Chairman to the April 1992 Participants' Meeting. Part Two: Work Program Fiscal Year 1992—Third Tranche*, Washington DC, GEF.

Global Environment Facility (1992c), *The Pilot Phase and Beyond*, Washington DC, Working Paper 1 Global Environment Facility.

Global Environment Facility (1993), *Mobilising Private Capital against Global Warming: A Business Concept and Policy Issues*, Washington DC, Global Environment Facility.

Goldammer, J.G. (ed.) (1990), *Fire in the Tropical Biota: Ecosystem Processes and Global Challenges*, Berlin, Springer Verlag.

Grainger, A. (1983), 'Improving the monitoring of deforestation in the humid tropics', in S.L. Sutton, T.C. Whitmore and A.C. Chadwick (eds), *Tropical Rainforest: Ecology and Management*, Oxford, Basil Blackwell, pp. 387–395.

Grainger, A. (1993), *Controlling Tropical Deforestation*, London, Earthscan.

Grayson, A.J. (1989), *Carbon Dioxide, Global Warming and Forestry*, Research Note 146, Alice Holt, Forestry Commission Research Division.

Grigal, D.F. and Ohmann, L.F. (1992), 'Carbon storage in upland forests of the Lake States', *Soil Science Society of America Journal*, **56**: 935–943.

Grubb, M. (1989), *The Greenhouse Effect: Negotiating Targets*, London, Royal Institute of International Affairs.

Grubb, M. (1993), 'Tradeable permits and the comprehensive approach to climate change: can we get the best of both worlds?', *Natural Resources Forum*, **17**: 51–57.

Grubb, M., Sebenius, J., Magalhaes, A. and Subak, S. (1992), 'Sharing the burden', in A.M. Mintzer (ed.), *Confronting Climate Change: Risks, Implications and Responses*, Cambridge, Cambridge University Press, pp. 305–322.

Grubb, M., Koch, M., Munson, A., Sullivan, F. and Thomson, K. (1993), *The 'Earth Summit' Agreements: A Guide and Assessment*, London, Earthscan.

Grübler, A., Nilsson, S. and Nakićenović, N. (1993), 'Enhancing carbon sinks', *Energy*, **18**: 499–522.

Gullison, R.E. and Losos, E.C. (1993), 'The role of foreign debt in deforestation in Latin America', *Conservation Biology*, **7**: 140–147.

Hahn, R.W. (1989), 'Economic prescriptions for environmental problems: how the patient followed the doctor's orders', *Journal of Economic Perspectives*, **3**: 95–114.

Haines, A. (1993), 'The possible effects of climate change on health', in E. Chivian, M. McCally, H. Hu and A. Haines (eds), *Critical Condition: Human Health and the Environment*, Cambridge, MA, MIT Press, pp. 151–170.

Hall, D.O. and Rosillo-Calle, F. (1990), 'African forests and grasslands: sources or sinks of greenhouse gases?', in S.H. Ominde and C. Juma (eds), *A Change in the Weather: African Perspectives on Climate Change*, Nairobi, ACTS Press, pp. 49–60.

Hall, D.O., Mynick, R.H. and Williams, R.H. (1990), *Carbon Sequestration versus Fossil Fuel Substitution*, Report No. 255, Princetown University, Center for Energy and Environmental Studies.

Hamilton, G.J. and Christie, J.M. (1971), *Forestry Management Tables*, Booklet 34, Edinburgh, Forestry Commission.

Hammond, A.L., Rodenburg, E. and Moomaw, W.R. (1990), 'Accountability in the greenhouse', *Nature*, **347**: 705–706.

Hanley, N. (ed.) (1991), *Farming and the Countryside: An Economic Analysis of External Costs and Benefits*, Wallingford, CAB International.

Hao, W.M., Liu, M.H. and Crutzen, P.J. (1990), 'Estimates of annual and regional releases of CO_2 and other trace gases to the atmosphere from fires in the tropics based on the FAO statistics for the period 1975–1980', in J.G. Goldammer (ed.), *Fire in the Tropical Biota*, Berlin, Springer Verlag, pp. 440–462.

Happell, J.D. and Chanton, J.P. (1993), 'Carbon remineralisation in a North Florida swamp forest: effects of water level on the pathways and rates of soil organic matter decomposition', *Global Biogeochemical Cycles*, **7**: 475–490.

Hardin, G. (1968), 'The tragedy of the commons', *Science*, **162**: 1243–1248.

Harmon, M.E., Ferrell, W.K. and Franklin, J.F. (1990), 'Effects on carbon storage of conversion of old-growth forests to young forests', *Science*, **247**: 699–702.

Harrison, P. (1992), *The Third Revolution: Environment, Population and a Sustainable World*, London, I.B. Tauris.

Harvey, D.R. (1990), 'The economics of the farmland market', in P.J. Dawson (ed.), *The Agricultural Land Market: Proceedings of an Agricultural Economics Society Conference*, University of Newcastle upon Tyne, Department of Agricultural Economics and Food Marketing, pp. 30–47.

Harvey, D.R. (1991), 'Agriculture and the environment: the way ahead', in N. Hanley (ed.), *Farming and the Countryside: An Economic Analysis of External Costs and Benefits*, Wallingford, CAB International, pp. 275–321.

Harvey, L.D.D. (1993), 'A guide to global warming potentials (GWPs)', *Energy Policy*, **21**: 24–34.

Hayes, P. and Smith, K.R. (1993a), 'Introduction', in P. Hayes and K.R. Smith (eds), *The Global Greenhouse Regime. Who Pays?*, London, Earthscan, pp. 3–19.

Hayes, P. and Smith, K.R. (eds) (1993b), *The Global Greenhouse Regime: Who Pays?*, London, Earthscan.

Hayes, T.D., Jewell, W.J., Dell'orto, S., Fanfoni, K.J., Leuschner, A.D. and Sherman, D.F. (1980), 'Anaerobic digestion of cattle manures', in D.A. Stafford, D.I. Wheatley and D.E. Hughes (eds), *Anaerobic Digestion*, London, Applied Science Publishers.

Heathwaite, A.L. (1993), 'Disappearing peat—regenerating peat? The impact of climate change on British peatlands', *Geographical Journal*, **159**: 203–208.

Hecht, S. (1993), 'Brazil: landlessness, land speculation and pasture-led deforestation', in M. Colchester and L. Lohmann (eds), *The Struggle for the Land and the Fate of the Forests*, London, Zed Books, pp. 164–178.

Hesse, P.R. (1984), 'Potential of organic materials for soil improvement', in IRRI (ed.), *Organic Matter and Rice*, Los Baños, Philippines, International Rice Research Institute, pp. 38–42.

Hewitt, K. (ed.) (1984), *Interpretations of Calamity*, London, Unwin Hyman.

Hodge, C.A.H., Buston, R.G.O., Corbett, W.M., Evans, R. and Seale, R.S. (1984), *Soils and their Use in Eastern England*, Harpenden, Soil Survey of England and Wales.

Holzapfel-Pschorn, A. and Seiler, W. (1986), 'Methane emissions during a cultivation period from an Italian rice paddy', *Journal of Geophysical Research*, **91**: 11803–11814.

Houghton, J.T., Callander, B.A. and Varney, S.K. (eds) (1992), *Climate Change 1992: The Supplementary Report to the IPCC Scientific Assessment*, Cambridge, Cambridge University Press.

Houghton, J.T., Jenkins, G.J. and Ephraums, J.J. (eds) (1990), *Climate Change: The IPCC Scientific Assessment*, Cambridge, Cambridge University Press.

Houghton, R.A. (1990), 'The future role of tropical forests in affecting the carbon dioxide concentration of the atmosphere', *Ambio*, **19**: 204–209.

Houghton, R.A. (1991), 'Tropical deforestation and atmospheric carbon dioxide', *Climatic Change*, **19**: 99–118.

Houghton, R.A. (1993a), 'Is carbon accumulating in the northern temperate zone?', *Global Biogeochemical Cycles*, **7**: 611–617.

Houghton, R.A. (1993b), 'The role of the world's forests in global warming', in K. Ramakrishna and G.M. Woodwell (eds), *World Forests for the Future: Their Use and Conservation*, New Haven, Yale University Press, pp. 21–58.

Houghton, R.A. and Skole, D.L. (1990), 'Carbon', in B.L. Turner, W.C. Clark, R.W. Kates, J.F. Richards, J.T. Mathews and W.B. Meyer (eds), *The Earth as Transformed by Human Action*, Cambridge, Cambridge University Press, pp. 393–408.

Houghton, R.A., Boone, R.D., Melillo, J.M., Palm, C.A., Woodwell, G.M., Myers, N., Moore, B. and Skole, D.L. (1985), 'Net flux of CO_2 from tropical forests in 1980', *Nature*, **316**: 617–620.

Houghton, R.A., Boone, R.D., Fruci, J.R., Hobbie, J.E., Melillo, J.M., Palm, C.A., Peterson, B.J., Shaver, G.R., Woodwell, G.M., Moore, B., Skole, D.L. and Myers, N. (1987), 'The flux of carbon from terrestrial ecosystems to the atmosphere in 1980 due to changes in land use', *Tellus*, **39B**: 122–139.

Houghton, R.A., Skole, D.L. and Lefkowitz, D.S. (1991), 'Changes in the landscape of Latin America between 1850 and 1985: 2 Net release of CO_2 to the atmosphere', *Forest Ecology and Management*, **38**: 173–199.

Houghton, R.A., Unruh, J.D. and Lefebvre, P.A. (1993), 'Current land cover in the tropics and its potential for sequestering carbon', *Global Biogeochemical Cycles*, 7: 305–320.

House of Commons Environment Committee (1993), *Forestry and the Environment*, Volume 1, London, HMSO.

Howarth, R.B. and Monahan, P.A. (1993), *Economics, Ethics and Climate Policy*, Stockholm, Stockholm Environment Institute.

Huke, R.E. (1982), *Rice Area by Type of Culture: South, South-East and East Asia*, Los Baños, Philippines, International Rice Research Institute.

Hulme, M. (1993a), 'Global warming', *Progress in Physical Geography*, **17**: 81–91.

Hulme, M. (1993b), 'Historic records and recent climatic change', in N. Roberts (ed.), *The Changing Global Environment*, Oxford, Blackwell, pp. 69–98.

Hulme, M., Hossell, J.E. and Parry, M.L. (1993), 'Future climate change and land use in the United Kingdom', *Geographical Journal*, **159**: 131–147.

Hunting Surveys and Consultants Limited (1986), *Monitoring Landscape Change*, London, HMSO.

Hyder, T.O. (1992), 'Climate negotiations: the North/South perspective', in A.M. Mintzer (ed.), *Confronting Climate Change: Risks, Implications and Responses*, Cambridge, Cambridge University Press, pp. 323–336.

Insley, H. (1988), *Farm Woodland Planning*, London, HMSO.

IPCC Response Strategies Working Group (1992), *Global Climate Change and the Rising Challenge of the Sea*, Netherlands, Report to UNEP, Ministry of Transport.

Isaksen, I.S.A., Ramaswamy, V., Rodhe, H. and Wigley, T.M.L. (1992), 'Radiative forcing of climate', in J.T. Houghton, B.A. Callander and S.K. Varney (eds), *Climate Change 1992: The Supplementary Report to the IPCC Scientific Assessment*, Cambridge, Cambridge University Press, pp. 47–67.

Ives, J.D. and Masserli, B. (1989), *The Himalayan Dilemma: Reconciling Development and Conservation*, London, Routledge.

Jenkinson, D.S. (1971), *The Accumulation of Organic Matter on Soil Left Uncultivated*, Harpenden, Rothamsted Experimental Station.

Jenkinson, D.S. (1988), 'Soil organic matter and its dynamics', in A. Wild (ed.), *Russell's Soil Conditions and Plant Growth*, London, Longman.

Johnson, B. (1991), *Responding to Tropical Deforestation*, Godalming, Worldwide Fund for Nature UK.

Jordan, A. (1994), 'Financing the UNCED agenda: the controversy over additionality', *Environment*, **36** (April): 16–20, 26–34.

Jordan, C.F. (ed.) (1989), *An Amazonian Rainforest: The Structure and Function of a Nutrient Stressed Ecosystem and the Impact of Slash and Burn Agriculture*, UNESCO Man and Biosphere, Volume 2, Paris, UNESCO.

Josling, T. and Tangerman, S. (1989), 'Measuring levels of protection in agriculture: a survey of approaches and results', in A. Maunder and A. Valdez (eds), *Agriculture and Governments in an Interdependent World: Proceedings of the Twentieth International Conference of Agricultural Economists*, Aldershot, Dartmouth Publishing Company, pp. 343–352.

Kahn, J.R. and MacDonald, J.A. (1992), *Third World Debt and Tropical Deforestation*, mimeo, Department of Economics, SUNY-Binghampton, New York.

Kahn, J. and MacDonald, J. (1994), 'International debt and deforestation', in K. Brown and D.W. Pearce (eds), *The Causes of Tropical Deforestation*, London, UCL Press, pp. 57–67.

Kane, S., Reilly, J. and Tobey, J. (1992), 'An empirical study of the economic effects of climate change on world agriculture', *Climatic Change*, **21**: 17–35.

Kasperson, R.E., Dow, K., Golding, D. and Kasperson, J.X. (eds) (1990), *Understanding Global Environmental Change: The Contributions of Risk Analysis and Management*, Worcester, MA, Center for Technology Environment and Development, Clark University.

Kates, R.W. and Haarmann, V. (1992), 'Where the poor live: are the assumptions correct?', *Environment*, **34** (4): 4–11, 25–28.

Katila, M. (1992), *Modelling Deforestation in Thailand: The Causes of Deforestation and Deforestation Projections for 1990–2010*, mimeo, Finnish Forestry Institute, Helsinki.

Kauppi, P.E., Mielikainen, K. and Kuusela, K. (1992), 'Biomass and carbon budget of European forests: 1971 to 1990', *Science*, 256: 70–74.

Khalil, M.A.K. and Rasmussen, R.A. (1990a), 'Atmospheric methane: recent global trends', *Environmental Science and Technology*, 24: 549–553.

Khalil, M.A.K. and Rasmussen, R.A. (1990b), 'Constraints on the global sources of methane and an analysis of recent budgets', *Tellus*, **42B**: 229–236.

Khalil, M.A.K., Rasmussen, R.A., Wang, M.X. and Ren, L.X. (1991), 'Methane emissions from rice fields in China', *Environmental Science and Technology*, 25: 979–981.

Knerr, B. (1992), 'Agricultural policies and deforestation in sub-Saharan Africa', in M. Loseby (ed.), *The Environment and the Management of Agricultural Resources: Proceedings of the Seminar of European Association of Agricultural Economists*, Viterbo, Italy, Università della Tuscia, pp. 70–90.

Korotkov, A.V. and Peck, T.J. (1993), 'Forest resources of the industrialised countries: an ECE/FAO assessment', *Unasylva*, **44**: 20–30.

Kramer, R.A. and Shabman, L. (1993), 'The effects of agricultural and tax policy reform on the economic return to wetland drainage in the Mississippi Delta region', *Land Economics*, **69**: 249–262.

Krugman, P.R. (1992), *Currencies and Crises*, Cambridge, MA, MIT Press.

Kummer, D. and Sham, C.H. (1994), 'The causes of tropical deforestation: a quantitative analysis and case study from the Philippines', in K. Brown and D.W. Pearce (eds), *The Causes of Tropical Deforestation*, London, UCL Press, pp. 146–158.

Kyrklund, B. (1990), 'The potential of trees and the forestry industry in reducing excess carbon dioxide', *Unasylva*, **163**: 12–14.

Lashof, D.A. (1989), 'The dynamic greenhouse: feedback processes that may influence future concentrations of atmospheric trace gases and climatic change', *Climatic Change*, 14: 213–242.

Ledec, G. (1992), *The Role of Bank Credit for Cattle Raising in Financing Tropical Deforestation: An Economic Case Study from Panama*, PhD Thesis, University of California at Berkeley.

Ledec, G. (1985), 'The political economy of tropical deforestation', in H.J. Leonard (ed.), *Divesting Nature's Capital: The Political Economy of Environmental Abuse in the Third World*, New York, Holmes and Meier, pp. 179–226.

Leggett, J. (1993), 'Who will underwrite the hurricane?', *New Scientist*, **139** (7 August): 29–33.

Lelieveld, J. and Crutzen, P.J. (1992), 'Indirect chemical effects of methane on climate warming', *Nature*, **355**: 339–342.

Lerner, J., Matthews, E. and Fung, I. (1988), 'Methane emission from animals: a global high-resolution database', *Global Biogeochemical Cycles*, **2**: 139–156.

Levine, J.S. (1990), 'Atmospheric trace gases: burning trees and bridges', *Nature*, **346**: 511–512.

Locke, G.M.L. (1987), *Census of Woodlands and Trees 1979–82*, Forestry Commission Bulletin 63, Edinburgh, Forestry Commission.

Lopez, R. (1992), 'Environmental degradation and economic openness in LDCs: the poverty linkage', *American Journal of Agricultural Economics*, **74**: 1138–1143.

Loske, R. and Oberthür, S. (1994), 'Joint implementation under the Climate Change Convention: opportunities and pitfalls', *International Environmental Affairs*, **6**: 45–58.

Lovins, A.B. and Lovins, L.H. (1992), 'Least cost climatic stabilisation', in G.I. Pearman (ed.), *Limiting Greenhouse Effects: Controlling Carbon Dioxide Emissions*, Chichester, John Wiley, pp. 351–442.

Lugo, A.E., Schmidt, R. and Brown, S. (1981), 'Tropical forests in the Caribbean', *Ambio*, **10**: 318–324.

Macmillan, D.C. (1993), 'Commercial forests in Scotland: an economic appraisal of replanting', *Journal of Agricultural Economics*, **44**: 51–66.

Maltby, E. (1986), *Waterlogged Wealth: Why Waste the World's Wet Places*, London, Earthscan.

Maltby, E. and Immirzi, C.P. (1993), 'Carbon dynamics in peatlands and other wetland soils: regional and global perspectives', *Chemosphere*, **27**: 999–1023.

Maltby, E., Immirzi, C.P. and McLaren, D.P. (1992), *Do Not Disturb: Peatbogs and the Greenhouse Effect*, London, Friends of the Earth.

Markandya, A. and Pearce, D.W. (1991), 'Development, the environment and the social rate of discount', *World Bank Research Observer*, **6**: 137–152.

Marland, G. (1988), *The Prospect of Solving the CO_2 Problem through Global Reforestation*, Report No. TR039, Washington DC, US Department of Energy Carbon Dioxide Research Division.

Mather, A.S. (1990), *Global Forest Resources*, London, Belhaven.

Mather, A. (1993), 'Afforestation in Britain', in A. Mather (ed.), *Afforestation: Policies, Planning and Progress*, London, Belhaven, pp. 13–33.

Mather, P.M. (1992), 'Remote sensing and the detection of change', in M.C. Whitby (ed.), *Land Use Change: The Causes and Consequences*, London, ITE Symposium No. 27, HMSO, pp. 60–68.

Matthews, E. (1983), 'Global vegetation and land use: new high resolution databases for climate studies', *Journal of Climate and Applied Meteorology*, **22**: 474–487.

Matthews, E. and Fung, I. (1987), 'Methane emissions from natural wetlands: global distribution and environmental characteristics of source', *Global Biogeochemical Cycles*, **1**: 61–86.

Matthews, E., Fung, I. and Lerner, J. (1991), 'Methane emission from rice cultivation: geographic and seasonal distribution of cultivated areas and emissions', *Global Biogeochemical Cycles*, **5**: 3–24.

McKibben, B. (1990), *The End of Nature*, London, Penguin.

Meat and Livestock Commission (1992), *CAP Reform: The Challenge of Change*, Milton Keynes, Meat and Livestock Commission.

Melillo, J.M., McGuire, A.D., Kicklighter, D.W., Moore, B., Vorosmarty, C.J. and Schloss, A.L. (1993), 'Global climate change and terrestrial net primary production', *Nature*, **363**: 234–240.

Ministry of Agriculture Fisheries and Food (1993a and various years), *Agriculture in the United Kingdom*, London, HMSO.

Ministry of Agriculture Fisheries and Food (1993b), *CAP Reform: Arable Area Payments 1993–1994: Explanatory Guide*, London, MAFF.

Molofsky, J., Menges, E.S., Hall, C.A.S., Armentano, T.V. and Ault, K.A. (1984), 'The effect of land use alteration on tropical carbon exhange', in T.N. Veziraglu (ed.), *The Biosphere: Problems and Solutions*, Amsterdam, Elsevier, pp. 181–184.

Moore, T.R. and Knowles, R. (1989), 'The influence of watertable levels on methane and carbon dioxide emissions from peatland soils', *Canadian Journal of Soil Science*, **69**: 33–38.

Mudge, F. and Adger, W.N. (1994), *Methane Emissions from Rice and Coarse Fibre Production*, Global Environmental Change Working Paper 94–08, Centre for Social and Economic Research on the Global Environment, University of East Anglia, University College London.

Myers, N. (1980), *Conversion of Moist Tropical Forests*, Washington DC, US National Research Council.

Myers, N. (1981), 'The hamburger connection: how Central America's forests become North America's hamburgers', *Ambio*, **10**: 3–8.

Myers, N. (1984), *The Primary Source: Tropical Forests and Our Future*, New York, Norton.

Myers, N. (1989a), *Deforestation Rates in Tropical Forests and their Climatic Implications*, London, Friends of the Earth.

Myers, N. (1989b), 'The greenhouse effect: a tropical forestry response', *Biomass*, **18**: 73–78.

Myers, N. (1990), 'Tropical forests', in J. Leggett (ed.), *Global Warming: The Greenpeace Report*, Oxford, Oxford University Press, pp. 372–399.

Myers, N. (1994), 'Tropical deforestation: rates and patterns', in K. Brown and D.W. Pearce (eds), *The Causes of Tropical Deforestation*, London, UCL Press, pp. 27–40.

National Research Council (1993), *Sustainable Agriculture and the Environment in the Humid Tropics*, Washington DC, National Academy Press.

Nations, J.D. (1992), 'Xateros, chicleros, and pimenteros: harvesting renewable tropical forest resources in the Guatemalan Peten', in K.H. Redford and C. Padoch (eds), *Conservation of Neotropical Forests*, New York, Columbia University Press, pp. 208–219.

Nature Conservancy Council and Countryside Commission for Scotland (1988), *National Countryside Monitoring Scheme Scotland: Grampian*, Coupar Angus, Countryside Commission for Scotland.

Neue, H.J. (1993), 'Methane emissions from rice fields', *BioScience*, **43**: 466–474.

Neue, H.U., Becker-Heidmann, P. and Scharpenseel, H.W. (1990), 'Organic matter dynamics, soil properties and cultural practices in rice land and their relationship to methane production', in A.F. Bouwman (ed.), *Soils and the Greenhouse Effect*, Chichester, John Wiley, pp. 457–466.

Nix, J. (1993), *Farm Management Pocketbook*, 24th Edn, Ashford, Kent, Wye College.

Nordhaus, W.D. (1991), 'To slow or not to slow: the economics of the greenhouse effect', *Economic Journal*, **101**: 920–937.

Nordhaus, W.D. (1992), 'An optimal transition path for controlling greenhouse gases', *Science*, **258**: 1315–1319.

Nordhaus, W.D. (1993), 'Optimal greenhouse gas reductions and tax policy in the DICE model', *American Economic Review: Papers and Proceedings*, **83**: 313–317.

OECD (1993), *Agricultural Policies, Markets and Trade: Monitoring and Outlook*, Paris, OECD.

O'Riordan, T. and Turner, R.K. (1983), 'The commons theme: introductory essay', in T. O'Riordan and R.K. Turner (eds), *An Annotated Reader in Environmental Planning and Management*, Oxford, Pergamon Press, pp. 265–288.

Overseas Development Administration (1988), *Appraisal of Projects in Developing Countries: A Guide for Economists*, London, HMSO.

Palo, M. (1990), 'Deforestation and development in the Third World: roles of system causality and population', in M. Palo and G. Mery (eds), *Deforestation or Development in the Third World?* Vol. III, Helsinki, Metsantutkimuslaitoksen Tiedonantoja (Research Bulletins of the Finnish Forest Research Institute), pp. 155–172.

Palo, M., Salmi, J. and Mery, G. (1987), 'Deforestation in the tropics: pilot scenarios based on quantitative analyses', in M. Palo and J. Salmi (eds), *Deforestation or Development in the Third World?* Vol. I, Helsinki, Metsantutkimuslaitoksen Tiedonantoja (Research Bulletins of The Finnish Forestry Institute), pp. 53–106.

Panayotou, T. and Ashton, P.S. (1993), *Not by Timber Alone: Economics and Ecology for Sustaining Tropical Forests*, Washington DC, Island Press.

Panayotou, T. and Sungsuwan, S. (1994), 'An econometric study of the causes of tropical deforestation: the case of northeast Thailand', in K. Brown and D.W. Pearce (eds), *The Causes of Tropical Deforestation*, London, UCL Press, pp. 192–210.

Parikh, J.K. (1992), 'IPCC strategies unfair to the South', *Nature*, **360**: 507–508.

Parry, M.L., Hossell, J.E. and Wright, L.J. (1992), 'Land use in the United Kingdom', in M.C. Whitby (ed.), *Land Use Change: the Causes and Consequences*, London, ITE Symposium No. 27, HMSO, pp. 7–14.

Pearce, D.W. (1991a), 'An economic approach to saving the tropical forest', in D. Helm (ed.), *Economic Policy Towards the Environment*, Oxford, Basil Blackwell, pp. 239–262.

Pearce, D.W. (1991b), 'Assessing the returns to the economy and society', in Forestry Commission (ed.), *Forestry Expansion: a Study of Technical, Economic and Ecological Factors*, Edinburgh, Forestry Commission.

Pearce, D.W. (1992), *The Secondary Benefits of Greenhouse Gas Control*, Global Environmental Change Working Paper 92–12, Centre for Social and Economic Research on the Global Environment, University College London and University of East Anglia.

Pearce, D.W. and Brown, K. (1994), 'Saving the world's tropical forests', in K. Brown and D.W. Pearce (eds), *The Causes of Tropical Deforestation*, London, UCL Press, pp. 2–26.

Pearce, D.W. and Puroshothaman, S. (1992), *Protecting Biological Diversity: The Economic Value of Pharmaceutical Plants*, Global Environmental Change Working Paper 92-27, Centre for Social and Economic Research on the Global Environment, University College London and University of East Anglia.

Pearce, D.W. and Warford, J. (1993), *World Without End: Economics, Environment and Sustainable Development*, Oxford, Oxford University Press.

Pearce, D.W., Maddison, D., Adger, W.N. and Moran, D. (1994), 'Foreign debt and environmental degradation: is there a link?', *Scientific American*, in press.

Pearce, F. (1989), *Turning Up the Heat: Our Perilous Future in the Global Greenhouse*, London, Paladin.

Peck, S.C. and Teisberg, T.J. (1992), 'CETA: a model for carbon emissions trajectory assessment', *Energy Journal*, **13**: 55–77.

Peluso, N.L. (1992), 'The political ecology of extraction and extractive reserves in East Kalimantan, Indonesia', *Development and Change*, **23**: 49–74.

Perrings, C. (1991), 'Reserved rationality and the precautionary principle: technological change, time and uncertainty in environmental decision-making', in R. Costanza (ed.), *Ecological Economics: The Science and Management of Sustainability*, New York, Columbia University Press, pp. 153–166.

Peters, C.M., Gentry, A.H. and Mendelsohn, R.O. (1989), 'Valuation of an Amazonian rainforest', *Nature*, **339**: 655–656.

Peters, R.L. (1988), 'The effect of global climatic change on natural communities', in E.O. Wilson (ed.), *Biodiversity*, Washington DC, National Academy Press, pp. 450–461.

Peters, R.L. (1990), 'Effects of global warming on forests', *Forest Ecology and Management*, **35**: 13–33.

Plotkin, M. and Famolare, L. (eds) (1992), *Sustainable Harvesting and Management of Rainforest Products*, Washington DC, Conservation International.

Ponnamperuma, F.N. (1984), 'Effects of drainage on the soil characteristics of paddy soils in China: comment on Cheng', in IRRI (eds), *Organic Matter and Rice*, Los Baños, Philippines, International Rice Research Institute, p. 427.

Porter, R.C. (1982), 'The new approach to wilderness preservation through benefit–cost analysis', *Journal of Environmental Economics and Management*, **9**: 59–80.

Poterba, J.M. (1993), 'Global warming policy: a public finance perspective', *Journal of Economic Perspectives*, **7**: 47–63.

Potter, C. (1991), 'Land diversion programmes as generators of public goods', in N. Hanley (ed.), *Farming and the Countryside: An Economic Analysis of External Costs and Benefits*, Wallingford, CAB International, pp. 230–249.

Price, C. (1993), *Time, Discounting and Value*, Oxford, Blackwell.

Price, C. and Willis, R. (1993), 'Time, discounting and the valuation of forestry's carbon fluxes', *Commonwealth Forestry Review*, **72**: 265–271.

Purseglove, J.W. (1974), *Tropical Crops: Dicotyledons*, London, Longman.

Purseglove, J.W. (1975), *Tropical Crops: Monocotyledons*, London, Longman.

Ramaswamy, V., Schwarzkopf, M.D. and Shine, K.P. (1992), 'Radiative forcing of climate from halocarbon-induced global stratospheric ozone loss', *Nature*, **355**: 810–812.

Rawcliffe, P. (1992), 'Lessons from the bogs: what now for the peat campaign?', *ECOS*, **13** (2): 41–47.

Redford, K.H. and Padoch, C. (eds) (1992), *Conservation of Neotropical Forests: Working from Traditional Resource Use*, New York, Columbia University Press.

Reed, D. (ed.) (1992), *Structural Adjustment and the Environment*, London, Earthscan.

Reid, W.V. and Trexler, M.C. (1991), *Drowning the National Heritage: Climate Change and US Coastal Biodiversity*, Washington DC, World Resources Institute.

Reid, W.V., Laird, S.A., Meyer, C.A., Gamez, R., Sittenfeld, A., Janzen, D., Gollin, M. and Juma, C. (1993), *Biodiversity Prospecting: Using Genetic Resources for Sustainable Development*, Washington DC, World Resources Institute.

Reis, E. and Guzman, R. (1994), 'An econometric model of Amazon deforestation', in K. Brown and D.W. Pearce (eds), *The Causes of Tropical Deforestation*, London, UCL Press, pp. 172–191.

Repetto, R. and Holmes, T. (1983), 'The role of population in resource depletion in developing countries', *Population and Development Review*, **9**: 607–632.

Repetto, R., Magrath, W., Wells, M., Beer, C. and Rossini, F. (1989), *Wasting Assets: Natural Resources in the National Income Accounts*, Washington DC, World Resources Institute.

Richards, J.F. (1990), 'Land transformation', in B.L. Turner, W.C. Clark, R.W. Kates, J.F. Richards, J.T. Mathews and W.B. Meyer (eds), *The Earth as Transformed by Human Action*, Cambridge, Cambridge University Press, pp. 163–178.

Richards, P. (1992), 'Saving the rainforest? Contested futures in conservation', in S. Wallman (ed.), *Contemporary Futures: Perspectives from Social Anthropology*, ASA Monograph 30, London, Routledge, pp. 138–156.

Robinson, J.M. (1989), 'On uncertainty in the computation of global emissions from biomass burning', *Climatic Change*, **14**: 243–262.

Rodhe, H., Eriksson, H., Robertson, K. and Svensson, B.H. (1991), 'Sources and sinks of greenhouse gases in Sweden: a case study', *Ambio*, **20**: 143–145.

Root, T.L. and Schneider, S.H. (1993), 'Can large scale climatic models be linked with multiscale ecological studies', *Conservation Biology*, **7**: 256–270.

Rose, A. (1992), 'Equity considerations of tradeable carbon emission entitlements', in UNCTAD (eds), *Combating Global Warming: Study on a Global System of Tradeable Carbon Emission Entitlements*, New York, United Nations, pp. 55–83.

Rosenberg, N.J. and Scott, M.J. (1994), 'Implications of policies to prevent climate change for future food security', *Global Environmental Change*, **4**: 49–62.

Rosenzweig, C. and Parry, M.L. (1994), 'Potential impact of climate change on world food supply', *Nature*, **367**: 133–138.

Rosenzweig, C., Parry, M.L., Fischer, G. and Frohberg, K. (1993), *Climate Change and World Food Supply*, Research Report No. 3, Environmental Change Unit, University of Oxford.

Rudel, T. (1989), 'Population, development and tropical deforestation: a cross national study', *Rural Sociology*, **54**: 327–338.

Ruitenbeek, H.J. (1990), *Economic Analysis of Tropical Forest Initiatives: Examples from West Africa*, Godalming, World Wide Fund for Nature UK.

Safley, M.L. and Westerman, P.W. (1988), 'Biogas production from anaerobic lagoons', *Biological Wastes*, **23**: 181–193.

Sass, R.L., Fischer, F.M., Harcombe, P.A. and Turner, F.T. (1991), 'Methane production and emission in a Texan agricultural wetland', *Global Biogeochemical Cycles*, **4**: 47–68.

Sathaye, J. and Reddy, A. (1993), 'Integrating ecology and economy in India', in P. Hayes and K.R. Smith (eds), *The Global Greenhouse Regime: Who Pays?*, London, Earthscan, pp. 191–209.

Schlesinger, W.H. (1991), *Biogeochemistry: An Analysis of Global Change*, London, Academic Press.

Schneider, R. (1992), *Brazil: An Analysis of Environmental Problems in the Amazon*, Report No. 9104-BR, Washington DC, World Bank.

Schneider, R. (1993), *The Potential for Trade with the Amazon in Greenhouse Gas Reduction*, LATEN Dissemination Note 2, Washington DC, World Bank.

Schneider, S.H. (1992), 'Global climate change: ecosystem effects', *Interdisciplinary Science Reviews*, **17**: 142–148.

Schroeder, P. (1992), 'Carbon storage potential of short rotation tropical tree plantations', *Forest Ecology and Management*, **50**: 31–41.

Schütz, H., Holzapfel-Pschorn, A., Conrad, R., Rennenberg, H. and Seiler, W. (1989), 'A three year continuous record on the influence of day time, season and fertiliser treatment on the methane emission rates from an Italian rice paddy field', *Journal of Geophysical Research*, **94**: 16 405–16 416.

Schütz, H., Seiler, W. and Rennenberg, H. (1990), 'Soil and land use related sources and sinks of methane in the context of the global methane budget', in A.F. Bouwman (ed.), *Soils and the Greenhouse Effect*, Chichester, John Wiley, pp. 268–285.

Schwartzsman, S. (1989), 'Extractive reserves in the Amazon', in J.O. Browder (ed.), *Fragile Lands of Latin America: Strategies for Economic Development*, Boulder, CO, Westview Press, pp. 150–165.

Sebacher, D.I., Harriss, K.B., Bartlett, S.M., Sebacher, S.M. and Grice, S.S. (1986), 'Atmospheric methane sources: Alaskan tundra bogs, an alpine fen and a subarctic boreal marsh', *Tellus*, **38B**: 1–10.

Sedjo, R.A. (1989), 'Forests to offset the greenhouse effect', *Journal of Forestry*, **87 (7)**: 12–15.

Sedjo, R.A. (1992), 'Temperate forest ecosystems in the global carbon cycle', *Ambio*, **21**: 274–277.

Sedjo, R.A. and Solomon, A.M. (1989), 'Climate and forests', in N.J. Rosenberg, W.E. Easterling, P. Crosson and J. Darmstadter (eds), *Greenhouse Warming: Abatement and Adaptation*, Washington DC, Resources for the Future, pp. 105–119.

Setzer, A.W. and Pereira, M.C. (1991), 'Amazonia biomass burnings in 1987 and an estimate of their tropospheric emissions', *Ambio*, **20**: 19–22.

Shafik, N. (1994), 'Macroeconomic causes of deforestation: barking up the wrong tree?', in K. Brown and D.W. Pearce (eds), *The Causes of Tropical Deforestation*, London, UCL Press, pp. 86–95.

Shine, K.P., Derwent, R.G., Wuebbles, D.J. and Morcrette, J.-J. (1990), 'Radiative forcing of climate', in J.T. Houghton, G.J. Jenkins and J.J. Ephraums (eds), *Climate Change: The IPCC Scientific Assessment*, Cambridge, Cambridge University Press, pp. 45–68.

Shugart, H.H., Antonovsky, M.Y., Jarvis, P.G. and Sandford, A.P. (1986), 'CO_2, climatic change and forest ecosystems', in B. Bolin, B.R. Döös, J. Jäger and R.A. Warrick (eds), *The Greenhouse Effect, Climate Change and Ecosystems*, Scope 29, Chichester, John Wiley, pp. 475–521.

Silvola, J. (1986), 'Carbon dioxide dynamics in mires reclaimed for forestry in eastern Finland', *Annales Botanici Fennici*, **23**: 59–67.

Singh, K.D. (1993), 'The 1990 Tropical Forest Resources Assessment', *Unasylva*, **44**: 10–19.

Smit, B., Ludlow, L. and Brklacich, M. (1988), 'Implications of a global climatic warming for agriculture: a review and appraisal', *Journal of Environmental Quality*, **17**: 519–527.

Smith, K.R. (1993), 'The basics of greenhouse gas indices', in P. Hayes and K.R. Smith (eds), *The Global Greenhouse Regime: Who Pays?*, London, Earthscan, pp. 20–50.

Sommer, A. (1976), 'Attempt at an assessment of the world's tropical forests', *Unasylva*, **28**: 5–25.

Southgate, D. (1994), 'Tropical deforestation and agricultural development in Latin America', in K. Brown and D.W. Pearce (eds), *The Causes of Tropical Deforestation*, London, UCL Press, pp. 134–144.

Southgate, D., Sierra, R. and Brown, L. (1989), *The Causes of Tropical Deforestation in Ecuador: A Statistical Analysis*, LEEC Discussion Paper 89–09, London Environmental Economics Centre.

Southworth, F., Dale, V.H. and O'Neill, R.V. (1991), 'Contrasting patterns of land use in Rondônia, Brazil: simulating the effects on carbon release', *International Social Science Journal*, **43**: 681–698.

Stewart, F. (1992), 'Short term policies for long term development', in G.A. Cornia, R. van der Hoeven and T. Mkandawire (eds), *Africa's Recovery in the 1990s: from Stagnation and Adjustment to Human Development*, New York, St. Martin's Press, pp. 312–333.

Street-Perrott, F.A. (1992), 'Atmospheric methane: tropical wetland sources', *Nature*, **355**: 23–24.

Strutt, N. (1970), *Modern Farming and the Soil: Report of the Agricultural Advisory Council on Soil Fertility and Soil Structure*, London, HMSO.

Subak, S. (1994), *Emissions for the Climate Change Convention: a Taxonomy of Sources*, Global Environmental Change Working Paper 94–07, Centre for Social and Economic Research on the Global Environment, University of East Anglia, University College London.

Subak, S., Raskin, P. and Von Hippel, D. (1993), 'National greenhouse gas accounts: current anthropogenic sources and sinks', *Climatic Change*, **25**: 15–58.

Svensson, B.H., Lantsheer, J.C. and Rodhe, H. (1991), 'Sources and sinks of methane in Sweden', *Ambio*, **20**: 155–160.

Swanson, T. (1994), *The International Regulation of Extinction*, London, Macmillan.

Swart, R.J., Bouwman, A.F., Livier, J. and Van der Born, G.J. (1993), 'Inventory of greenhouse gas emissions in the Netherlands', *Ambio*, **22**: 518–523.

Swisher, J.N. (1991), 'Cost and performance of CO_2 storage in forestry projects', *Biomass and Bioenergy*, **1**: 317–328.

Tans, P.P., Fung, I.Y. and Takahashi, T. (1990), 'Observational constraints on the global atmospheric CO_2 budget', *Science*, **247**: 1431–1438.

Tate, K.R. (1992), 'Assessment, based on a climosequence of soils in tussock grasslands, of soil carbon storage and release in response to global warming', *Journal of Soil Science*, **43**: 697–707.

Tathy, J.P., Delmas, R.A., Marenco, A., Cros, B., Labat, M. and Servant, J. (1992), 'Methane emission from flooded forest in Central Africa', *Journal of Geophysical Research*, **97**: 6159–6168.

Taylor, J.A. (1983), 'The peatlands of Great Britain and Ireland', in A.J.P. Gore (ed.), *Ecosystems of the World Vol. 4A: Mires: Swamp, Bog, Fen and Moor*, Amsterdam, Elsevier, pp. 1–46.

Taylor, J. (1993), 'The mutable carbon sink', *Nature*, **366**: 515–516.

Thompson, D.A. and Matthews, R.W. (1989), *The Storage of Carbon in Trees and Timber*, Research Note 160, Forestry Commission Research Division, Alice Holt.

Tietenberg, T. (1992), 'Implementation issues: a general survey', in UNCTAD (ed.), *Combating Global Warming: Study on a Global System of Tradeable Carbon Emission Entitlements*, New York, United Nations, pp. 127–149.

Tiffen, M., Mortimore, M. and Gachuki, F. (1994), *More People, Less Erosion*, Chichester, John Wiley.

Trexler, M.C., Faeth, P.E. and Kramer, J.M. (1989), *Forestry as a Response to Global Warming: An Analysis of the Guatemala Agroforestry and Carbon Sequestration Project*, Washington DC, World Resources Institute.

Turner, B.L., Kasperson, R.E., Meyer, W.B., Dow, K.M., Golding, D., Kasperson, J.X., Mitchell, R.C. and Ratick, S.J. (1990), 'Two types of global environmental change:

definitional and spatial-scale issues in their human dimensions', *Global Environmental Change*, **1**: 14–22.

Turner, R.K. and Jones, T. (eds) (1991), *Wetlands: Market and Intervention Failures*, London, Earthscan.

Turner, R.K., Doktor, P. and Adger, W.N. (1993), 'Sea level rise vulnerability assessment of UK coastal zones', Global Environmental Change Working Paper 93–27, Centre for Social and Economic Research on the Global Environment, University of East Anglia.

Uhl, C. (1987), 'Factors controlling succession following slash and burn agriculture in Amazonia', *Journal of Ecology*, **75**: 377–407.

UNCTAD (United Nations Conference on Trade and Development) (1992), *Combating Global Warming: Study on a Global System of Tradeable Carbon Emission Entitlements*, New York, United Nations.

United Nations Environment Programme (1991), *Environmental Data Report 1991–1992*, Oxford, Blackwell.

Vanclay, J. (1992), 'Species richness and productive forest management', in F.R. Miller and K.L. Adam (eds), *Oxford Conference on Tropical Forests*, Oxford, Oxford Forestry Institute, pp. 1–10.

Vincent, J.R. (1992), 'The tropical timber trade and sustainable development', *Science*, **256**: 1651–1655.

Walsh, R.G. (1986), *Recreation Economic Decisions: Comparing Costs and Benefits*, State College, PA, Venture Publishing.

Warrick, R.A. (1993), 'Slowing global warming and sea-level rise: the rough road from Rio', *Transactions of the Institute of British Geographers*, **18**: 140–148.

Wassmann, R., Papen, H. and Rennenberg, H. (1993), 'Methane emissions from rice paddies and possible mitigation strategies', *Chemosphere*, **26**: 201–217.

Watanabe, I. (1984), 'Anaerobic decomposition of organic matter in flooded rice soils', in IRRI (eds), *Organic Matter and Rice*, Los Baños, Philippines, International Rice Research Institute, pp. 237–257.

Watson, R.T., Rodhe, H., Oeschger, H. and Siegenthaler, U. (1990), 'Greenhouse gases and aerosols', in J.T. Houghton, G.J. Jenkins and J.J. Ephraums (eds), *Climate Change: The IPCC Scientific Assessment*, Cambridge, Cambridge University Press, pp. 1–40.

Watson, R.T., Meira Filho, L.G., Sanhueza, E. and Janetos, A. (1992), 'Greenhouse gases: sources and sinks', in J.T. Houghton, B.A. Callander and S.K. Varney (eds), *Climate Change 1992: The Supplementary Report to the IPCC Scientific Assessment*, Cambridge, Cambridge University Press, pp. 29–46.

Westoby, J. (1989), *Introduction to World Forestry*, Oxford, Basil Blackwell.

Whiting, G.J. and Chanton, J.P. (1993), 'Primary production control of methane emission from wetlands', *Nature*, **364**: 794–795.

Wigley, T.M.L. (1991), 'A simple inverse carbon cycle model', *Global Biogeochemical Cycles*, **5**: 373–382.

Wigley, T.M.L. and Raper, S.C.B. (1992), 'Implications for climate and sea level rise of revised IPCC emission scenarios', *Nature*, **357**: 293–300.

Wigley, T.M.L. and Raper, S.C.B. (1993), 'Global mean temperature and sea level projections under the IPCC emissions scenarios', in R.A. Warrick, E.M. Barrow and T.M.L. Wigley (eds), *Climate and Sea Level Change: Observations, Projections and Implications*, Cambridge, Cambridge University Press, pp. 401–404.

Williams, C.N., Chew, W.Y. and Rajaratnam, J.A. (1980), *Tree and Field Crops of the Wetter Regions of the Tropics*, Harlow, Longman.

Willis, K.G., Benson, J.F. and Saunders, C.M. (1988), 'The impact of agricultural policy on the costs of nature conservation', *Land Economics*, **64**: 147–157.

Wilson, E.O. (ed.) (1988), *Biodiversity*, Washington DC, National Academy Press.

Wood, W.B. (1990), 'Tropical deforestation: balancing regional development demands and global environmental concerns', *Global Environmental Change*, **1**: 23–41.

Woodwell, G.M. (1992), 'The role of forests in climatic change', in N.P. Sharma (ed.), *Managing the World's Forests: Looking for Balance between Conservation and Development*, Dubuque, IA, Kendall Hunt, pp. 75–91.

World Bank (1992), *World Development Report 1992*, New York, Oxford University Press.

World Commission on Environment and Development (1987), *Our Common Future*, Oxford, Oxford University Press.

World Resources Institute (1990), *World Resources 1990–1991*, Oxford, Oxford University Press.

World Resources Institute (1992), *World Resources 1992–1993*, Oxford, Oxford University Press.

Wuebbles, D.J. and Edmonds, J. (1991), *Primer on Greenhouse Gases*, Chelsea, MI, Lewis Publishers.

Wuebbles, D.J., Patten, K.O., Grant, K.E. and Jain, A.K. (1992), *Sensitivity of Direct Global Warming Potentials to Key Uncertainties*, California, Lawrence Livermore National Laboratory.

Yagi, K. and Minami, K. (1990), 'Effects of organic matter applications on methane emission from Japanese paddy fields', in A.F. Bouwman (ed.), *Soils and the Greenhouse Effect*, Chichester, John Wiley, pp. 467–473.

Young, A. (1976), *Tropical Soils and Soil Survey*, Cambridge, Cambridge University Press.

Index